Certification Study Companion Series

The Apress Certification Study Companion Series offers guidance and hands-on practice to support technical and business professionals who are studying for an exam in the pursuit of an industry certification. Professionals worldwide seek to achieve certifications in order to advance in a career role, reinforce knowledge in a specific discipline, or to apply for or change jobs. This series focuses on the most widely taken certification exams in a given field. It is designed to be user friendly, tracking to topics as they appear in a given exam and work alongside other certification material as professionals prepare for their exam.

More information about this series at https://link.springer.com/bookseries/17100.

The Splunk Core User Study Companion

Achieve Splunk Enterprise Certified Admin and Gain Architect Essentials

Second Edition

Carlos Moreno Buitrago
Deep Mehta

Apress®

The Splunk Core User Study Companion: Achieve Splunk Enterprise Certified Admin and Gain Architect Essentials, Second Edition

Carlos Moreno Buitrago
Toronto, ON, Canada

Deep Mehta
Mumbai, Maharashtra, India

ISBN-13 (pbk): 979-8-8688-2500-2
https://doi.org/10.1007/979-8-8688-2501-9

ISBN-13 (electronic): 979-8-8688-2501-9

Managing Director, Apress Media LLC: Welmoed Spahr
Acquisitions Editor: James Robinson-Prior
Editorial Project Manager: Gryffin Winkler

Cover designed by eStudioCalamar

Distributed to the book trade worldwide by Springer Science+Business Media New York, 1 New York Plaza, New York, NY 10004. Phone 1-800-SPRINGER, fax (201) 348-4505, e-mail orders-ny@springer-sbm.com, or visit www.springeronline.com. Apress Media, LLC is a Delaware LLC and the sole member (owner) is Springer Science + Business Media Finance Inc (SSBM Finance Inc). SSBM Finance Inc is a Delaware corporation.

For information on translations, please e-mail booktranslations@springernature.com; for reprint, paperback, or audio rights, please e-mail bookpermissions@springernature.com.

Apress titles may be purchased in bulk for academic, corporate, or promotional use. eBook versions and licenses are also available for most titles. For more information, reference our Print and eBook Bulk Sales web page at http://www.apress.com/bulk-sales.

Any source code or other supplementary material referenced by the author in this book is available to readers on GitHub. For more detailed information, please visit https://www.apress.com/gp/services/source-code.

If disposing of this product, please recycle the paper.

*To my parents, for your constant support
and unconditional love.
To my brother and sister, for the laughter,
strength, and balance you bring to my life.
Part of who I am, and even what lives on these pages,
exists because of you.*

Table of Contents

Chapter 10: Advanced Data Input in Splunk 247

About the Authors

Carlos Moreno Buitrago is an observability and cybersecurity specialist focused on Splunk architecture, operations, and administration. He designs end-to-end data pipelines related to log, metric, and event flows, from source to search, with a strong emphasis on governance, reliability, and cost control. Carlos has deep hands-on experience with Cribl for routing, shaping, enrichment, and ROI optimization, alongside Splunk features like indexer/search head clustering, HEC, CIM/data models, and Enterprise Security content. Additionally, Carlos has worked across leading security and networking stacks, bridging security operations with platform engineering. That experience helps teams turn messy telemetry into trustworthy, searchable data that powers real-world detections and business insights.

Deep Mehta is an AWS Certified Associate Architect, Docker Certified Associate, Certified Splunk Architect (ongoing), and Certified Splunk User, Power User, and Admin. He's worked on the Splunk platform since 2017, having experience consulting in the telecommunication, aviation, and healthcare industries. Apart from being passionate about big data technologies, he also loves playing squash and badminton.

About the Technical Reviewer

 **Sivaraj Selvaraj's** work is focused on modern technologies and industry best practices. His experience includes front-end development approaches such as HTML5, CSS3, and JavaScript frameworks, as well as creating responsive web designs to optimize user experience across devices. He specializes in developing dynamic web applications with server-side languages such as PHP, WordPress, and Laravel, as well as managing and integrating databases with SQL and MySQL. Sivaraj is enthusiastic about sharing his significant expertise and experience, empowering readers to solve tough challenges and create highly functional, visually appealing websites.

Acknowledgments

Writing this book was as much a learning journey as it was a technical one. Along the way, we benefited from the knowledge and experience of many people, but above all, we are grateful to you, the reader, for choosing this book.

We greatly appreciate the official documentation and the Splunk community, whose accuracy, openness, and discussions have constantly inspired us and helped shape many of the ideas in this book.

We would also like to thank the colleagues and mentors who shared their experience, reviewed ideas, and encouraged us to keep moving forward. Your support, whether direct or indirect, played an important role in bringing this book to life.

Introduction

Splunk is a software technology for monitoring, searching, analyzing, and visualizing machine-generated data across modern IT environments. From collecting logs on a single server to running large, multisite, high-availability deployments in the cloud, Splunk is both powerful and complex. For many administrators, the challenge is not a lack of features; it is understanding how all the pieces fit together and how to operate them correctly in real-world scenarios. This book is designed to guide you through that journey.

It discusses the roles of a Splunk admin and explains how Splunk architecture can be efficiently deployed and managed. It covers everything you need to know to ace the Splunk exams or build and maintain Splunk environments in production. The book is written to be used interactively and includes practice datasets and test questions at the end of every chapter.

The content is structured to build progressively, starting with foundational concepts and moving toward advanced administration topics. There are three modules dedicated to achieving the Splunk Enterprise Certified Admin. The first module comprises seven chapters dedicated to passing the Splunk Core Certified User and the Splunk Core Certified Power User exams. It covers installing Splunk, Splunk's Search Processing Language (SPL), field extraction, field aliases and macros in Splunk, creating Splunk tags, Splunk lookups, and invoking alerts. You will learn how to make a data model and prepare an advanced dashboard in Splunk. The second module is dedicated to the Splunk Enterprise Certified Admin exam and consists of five chapters. It covers Splunk licenses and user role management, Splunk forwarder configuration, indexer clustering, Splunk security policies, and advanced data input options.

The third module focuses on teaching Splunk admins to troubleshoot
and manage the Splunk infrastructure, which is a part of the Splunk
Enterprise Certified Architect exam. Aditionally, you will learn how to set
up Splunk Enterprise on the AWS platform, and you will be introduced
to some of the best practices in Splunk. At the end of every chapter is
a multiple-choice test to help candidates become more familiar with
the exam.

PART I

Splunk Architecture, Splunk SPL (Search Processing Language), and Splunk Knowledge Objects

CHAPTER 1

An Overview of Splunk

Splunk is a data analytics and observability platform for logs, metrics, and traces. It ingests machine data from applications, infrastructure, cloud services, and security tools; indexes it at scale; and lets you interrogate it with SPL (Search Processing Language) using a schema-on-read model. Beyond ad hoc searching and monitoring, Splunk powers real-time detection, historical analytics, dashboards, and automated actions, with major use cases in SIEM and security analytics.

This chapter discusses the basics of Splunk, including its history and architecture, and delves into how to install the software on local machines. You see the layout of the Splunk Enterprise Certified Admin exam, learn how to add user data and a props.conf file, and know the process of editing timestamps, which is useful in the later chapters. A few sample exam questions are at the end of the chapter.

Summing it up, this chapter covers the following topics:

- An overview of the Splunk Enterprise Certified Admin exam

- An introduction to Splunk

- The Splunk architecture

- Installing Splunk on macOS and Windows

- Adding data to Splunk

C. M. Buitrago and D. Mehta, *The Splunk Core User Study Companion*, Certification Study Companion Series, https://doi.org/10.1007/979-8-8688-2501-9_1

Overview of the Splunk Admin Exam

A Splunk Enterprise Certified Admin is responsible for the daily management of Splunk Enterprise, including performance monitoring, security enhancement, license management, indexers and search heads, configuration, and adding data to Splunk. The following are the areas of expertise that the exam tests:

- Splunk deployment

- License management

- Splunk apps

- Splunk configuration files

- Users, roles, and authentication

- Get Data In (GDI)

- Distributed search

- Introduction to Splunk clusters

- Deploying forwarders with forwarder management

- Configuring common Splunk data inputs

- Customizing the input parsing process

In the next section, you will learn about the admin exam's structure.

Structure

The Splunk Enterprise Certified Admin exam is in multiple-choice question format. You have 57 minutes to answer 56 questions and an additional 3 minutes to review the exam agreement, totaling 60 minutes. The passing scores are not public-facing information, but should be higher than 70%.

The exam's registration fee is $130 (USD). Refer to `https://www.splunk.com/en_us/pdfs/training/splunk-test-blueprint-enterprise-admin.pdf` for more information.

The exam questions come in three formats:

- **Multiple choice**: You must select the option that is the best answer to a question or to complete a statement.

- **Multiple responses**: You must select the options that best answer a question or complete a statement.

- **Sample directions**: You read a statement or question and select only the answer(s) that represent the most ideal or correct response.

Requirements

The Splunk Enterprise Certified Admin exam has a prerequisite. You must first pass the following exam:

- The Splunk Core Certified Power User exam

The following is a suggested list of training from the Splunk Enterprise Certified Admin Learning Path that covers the areas of expertise mentioned before:

- Splunk Enterprise System Administration

- Splunk Enterprise Data Administration

Splunk offers multiple free courses on its website to prepare for the Splunk Core Certified User and the Splunk Core Certified Power User certifications. The Splunk Core Certified User exam is optional; it can be skipped. However, in this book, we will cover the content of both exams to prepare for the admin exam.

The learning flow is shown in Figure 1-1.

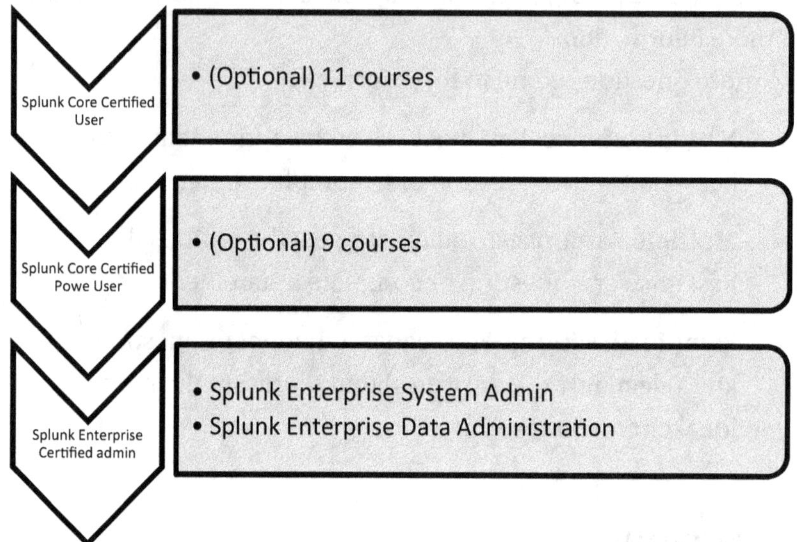

Figure 1-1. *Splunk exam prerequisites*

- **Splunk Courses (11, formerly Fundamentals 1):**
  Offered to students in two ways: e-learning or instructor-led. These courses introduce you to the Splunk platform.

- **Splunk Core Certified User Exam** tests your knowledge of and skills in searching, using fields, creating alerts, using lookups, and creating basic statistical reports and dashboards.

- **Splunk Courses (9, formerly Fundamentals 2):**
  Offered to students in two ways: e-learning or instructor-led. The courses introduce you to searching and reporting commands and creating knowledge objects.

- The **Splunk Core Certified Power User exam** tests the knowledge and skills required for SPL searching and reporting commands and building knowledge objects, using field aliases and calculated fields, creating tags and event types, using macros, creating workflow actions and data models, and normalizing data with the Common Information Model.

- **Splunk Enterprise System Administration** is instructor-led and designed for system administrators responsible for managing the Splunk Enterprise environment. The course teaches fundamental information for Splunk license managers, indexers, and search heads.

- **Splunk Enterprise Data Administration** is instructor-led and designed for system administrators responsible for adding remote data to Splunk indexers. The course provides fundamental information on Splunk forwarders and methods.

- The **Splunk Enterprise Certified Admin** exam tests your knowledge of and skills in managing various components of Splunk Enterprise, including license management, indexers and search heads, configuration, monitoring, and adding data to Splunk.

Modules 2 and 3 of this book focus on the Splunk Enterprise system administration and data administration exams.

Blueprint

The Splunk Enterprise Certified Admin exam has 17 sections, described as follows:

- **Section 1: Splunk Admin Basics (5%)**: This section focuses on identifying Splunk components.

- **Section 2: License Management (5%)**: This section focuses on identifying license types and understanding license violations.

- **Section 3: Splunk Configuration Files (5%)**: This section focuses on configuration layering, configuration precedence, and the btool command-line tool to examine configuration settings.

- **Section 4: Splunk Indexes (10%)**: This section focuses on basic index structure, types of index buckets, checking index data integrity, and the workings of the indexes.conf file, fishbucket and the data retention policy.

- **Section 5: Splunk User Management (5%)**: This section focuses on user roles, creating a custom role, and adding Splunk users.

- **Section 6: Splunk Authentication Management (5%)**: This section focuses on LDAP, user authentication options, and multifactor authentication.

- **Section 7: Getting Data In (5%)**: This section focuses on basic input settings, Splunk forwarder types, configuring the forwarder, and adding UF input using CLI.

- **Section 8: Distributed Search (10%)**: This section focuses on distributed search, the roles of the search head and search peers, configuring a distributed search group, and search head scaling options.

- **Section 9: Getting Data In—Staging (5%)**: This section focuses on the three phases of the Splunk indexing process and Splunk input options.

- **Section 10: Configuring Forwarders (5%)**: This section focuses on configuring forwarders and identifying additional forwarder options.

- **Section 11: Forwarder Management (10%)**: This section focuses on deployment management, the deployment server, managing forwarders using deployment apps, configuring deployment clients, configuring client groups, and monitoring forwarder management activities.

- **Section 12: Monitor Inputs (5%)**: This section examines your knowledge of file and directory monitor inputs, optional settings for monitor inputs, and deploying a remote monitor input.

- **Section 13: Network and Scripted Inputs (5%)**: This section examines your knowledge of the network (TCP and UDP) inputs, optional settings for network inputs, and a basic scripted input.

- **Section 14: Agentless Inputs (5%)**: This section examines your knowledge of Windows input types and the HTTP event collector.

- **Section 15: Fine-Tuning Inputs (5%)**: This section examines your knowledge of the default processing during the input phase and configuring input phase options, such as source type fine-tuning and character set encoding.

- **Section 16: Parsing Phase and Data (5%)**: This section examines your knowledge of the default processing during parsing, optimizing, and configuring event line breaking, extraction of timestamps and time zones from events, and data preview to validate events created during the parsing phase.

- **Section 17: Manipulating Raw Data (5%)**: This section examines your knowledge of how data transformations are defined and invoked, and the use of transformations with props.conf and transforms.conf and SEDCMD to modify raw data.

An Introduction to Splunk

The word *Splunk* comes from the word *spelunking*, which means to explore caves. Splunk can analyze almost all known data types, including machine data, structured data, and unstructured data. Splunk provides operational feedback on what is happening across an infrastructure in real time—facilitating fast decision-making.

Splunk is commonly thought of as "a Google for log files" because, like Google, you can use Splunk to determine the state of a network and the activities taking place within it. It is a **centralized log management tool**, but it also works well with structured and unstructured data. Splunk **monitors, reports, and analyzes real-time machine data** and indexes data based on timestamps.

The History of Splunk

Splunk was founded by Rob Das, Erik Swan, and Michael Baum in October 2003. It grew from a small startup company to one of the biggest multinational corporations for security information and event management (SIEM) tools. Before Splunk, a business needing to troubleshoot its environment had to rely on the IT department, where a programmer wrote scripts to meet needs. This script ran on top of a platform to generate a report.

As a result, companies didn't have a precise way to discover problems deep inside their infrastructure. Splunk was created to deal with this issue. Initially, Splunk focused on analyzing and understanding a problem, learning what organizations do when something goes wrong, and retracing footprints.

The first version of Splunk was released in 2004 in the Unix market, where it started to gain attention. Cisco Systems Inc. acquired Splunk for $157 per share in cash, representing approximately $28 billion in equity value on March 18, 2024.

It is important to understand why this software was developed. The following section discusses Splunk's many useful benefits.

The Benefits of Splunk

Splunk offers a variety of benefits, including the following:

- Converts complex log analysis report into graphs

- Supports structured as well as unstructured data

- Provides a simple and scalable platform

- Offers a simple architecture for the most complex architecture

- Understands machine data

- Provides data insights for operational intelligence

- Monitors IT data continuously

The Splunk Architecture

The Splunk indexer works in a specified manner in a set architecture (see Figure 1-2).

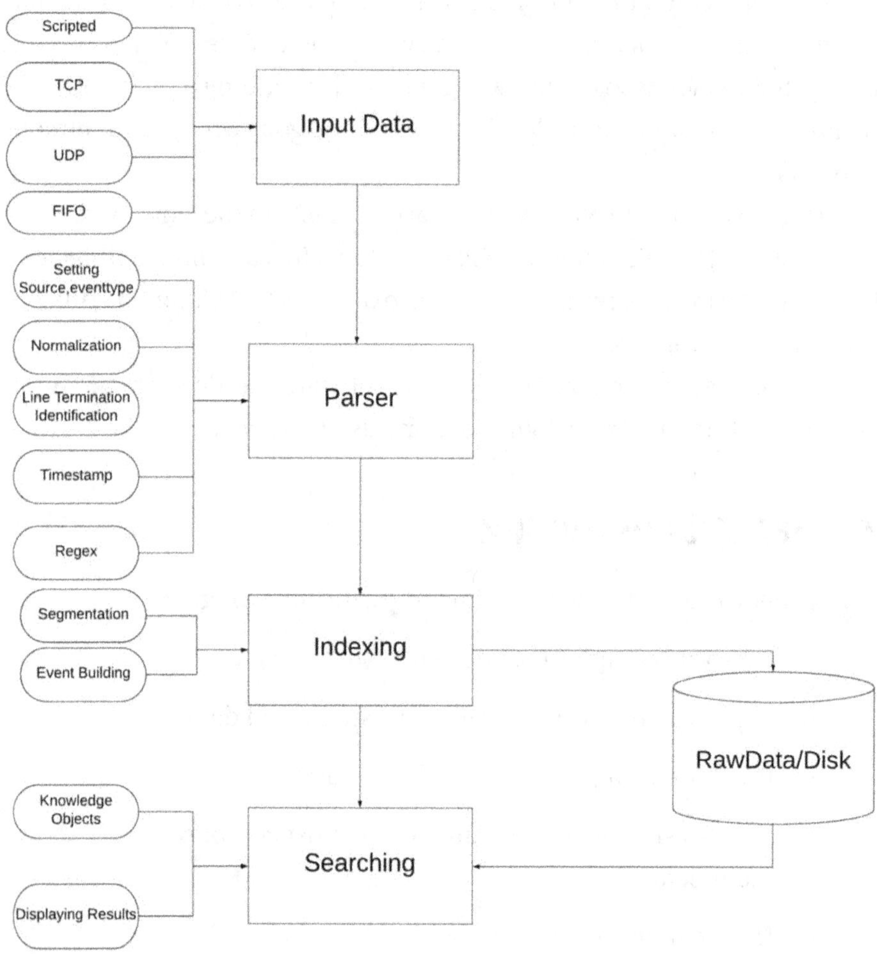

Figure 1-2. *Splunk architecture diagram*

Let's parse this diagram and introduce its components:

- **Input data**: This is the first phase of onboarding data. There are several methods to bring data into Splunk: it can listen to your ports, your REST API endpoint, the Transmission Control Protocol (TCP), the User Datagram Protocol (UDP), and so on, use scripted input, read a file, or the Windows Event Log.

- **Parser**: The second phase is to parse the input, in which a chunk of data is broken into various events. In the parsing phase, you can extract default fields, such as the source type. You can also extract timestamps from the data, identify the line's termination, and perform other similar actions. You can also mask sensitive but useful data. For example, if the data is from a bank and includes customers' account numbers, masking data is essential. In the parsing phase, you can apply custom metadata if required.

- **Indexing**: At this stage, the event is divided into segments within which the search can be performed. The data is written to disk, and you can design indexing data structures.

- **Searching**: In this phase, the search operations are performed on top of the index data, and you can create knowledge objects and perform queries, reports, dashboards, etc. For example, you are able to create a monthly sales report and send it weekly by email.

In a standalone machine, the input data, the parser, and the indexer are in the same instance. In contrast to this, in a distributed environment, the input data is parsed to the Indexer (IDX) or the Heavy Forwarder (HF) using Universal Forwarder (UF), which is a lightweight program that gets data into Splunk. In the UF, you cannot search for data or perform any operation. You look at this later in the chapter.

Figure 1-3 shows Splunk's architecture.

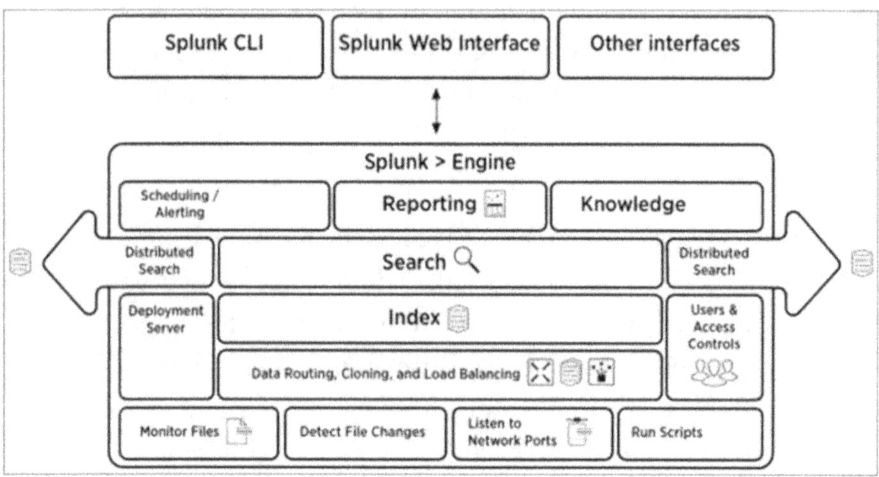

Figure 1-3. *Splunk's architecture*

The following tasks can be performed in the Splunk architecture:

- You can **receive data** through network ports, monitor files, and detect file changes in real time.

- You can run scripts to get **customized data**.

- Data routing, cloning, and load balancing are available in **distributed environments**, which you learn about in later chapters.

- **User access controls** preserve security. There are various default role levels: user, power user, and admin. Users can create knowledge objects or read indexes based on their rights.

- The deployment server distributes configurations and apps for the stack. You can deploy new applications using the **deployment server**.

- When an indexer receives data from a parser, it indexes it, and you can break down the event into **segments**.

- Once the data is stored in the indexer, you can perform **search operations**.

- You can do a scheduled search on indexed data.

- You can **generate a data alert** by setting parameters; for example, when the transaction time exceeds 15 minutes in a particular transaction.

- You can **create reports** by scheduling, saving searches, or creating a macro. There are a variety of ways to generate reports.

- Knowledge objects are useful for creating specialized reports from user-defined data, unstructured data, and so on.

- You can **access the Splunk instance** using either the Splunk Web interface, which is the most popular option, or the Splunk CLI.

Now, let's move forward to learn how to get Splunk quickly installed.

Installing Splunk

You can download and install Splunk Enterprise for free using its 60-day trial version that indexes 500 MB/day. All you need to do is create an account at `www.splunk.com/en_us/download/splunk-enterprise.html`.

After 60 days, you can convert to a perpetually free license or purchase a Splunk Enterprise license to continue using the expanded functionality designed for enterprise-scale deployments (Chapter 8 discusses Splunk licenses in detail).

Table 1-1 shows the relationship of a few Splunk attributes to their default ports during installation. The importance of each attribute is discussed later in the book.

Table 1-1. *Splunk attributes and default port values*

Attribute	Default Port
Splunk default port	8000
Splunk management port	8089
Splunk KV Store	8191

Installing Splunk on macOS

macOS users should follow these steps to install Splunk:

1. Sign in to your Splunk account.

2. Download the Splunk file at `https://www.splunk.com/en_us/download/splunk-enterprise.html#osx.`

3. Download the .dmg file.

4. Open the downloaded file.

5. Click the Install button.

6. Click the Continue button in the next step.

7. Click the Continue button until you are prompted to click the Install button.

8. If you want to change the install location first, you can do it by clicking Change Install Location (see Figure 1-4).

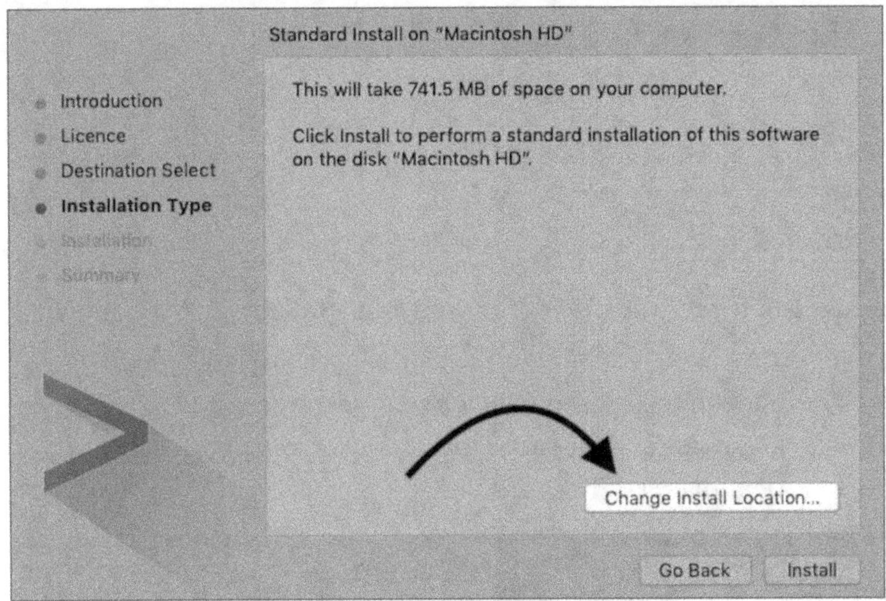

Figure 1-4. *Installing Splunk*

9. After selecting the path in which you want to install Splunk, click the Install button.

10. Depending on the version you installed, you will be asked to enter the user and password, or run the following command:

    ```
    /Applications/Splunk/bin/splunk start --accept-
    license
    ```

11. Enter **admin** as the administrator username (see Figure 1-5).

This appears to be your first time running this version of Splunk.

Splunk software must create an administrator account during startup. Otherwise, you cannot log in.
Create credentials for the administrator account.
Characters do not appear on the screen when you type in credentials.

Please enter an administrator username:

Figure 1-5. *Create administrator user*

12.   Enter your new password, as shown in Figure 1-6.

```
This appears to be your first time running this version of Splunk.

Splunk software must create an administrator account during startup. Otherwise, you cannot log in.
Create credentials for the administrator account.
Characters do not appear on the screen when you type in credentials.

Please enter an administrator username: admin
Password must contain at least:
   * 8 total printable ASCII character(s).
Please enter a new password:
```

Figure 1-6. *Create password for the administrator user*

13.   Start Splunk at **http://localhost:8000**. Enter the
username and password that you just created
(Figure 1-7).

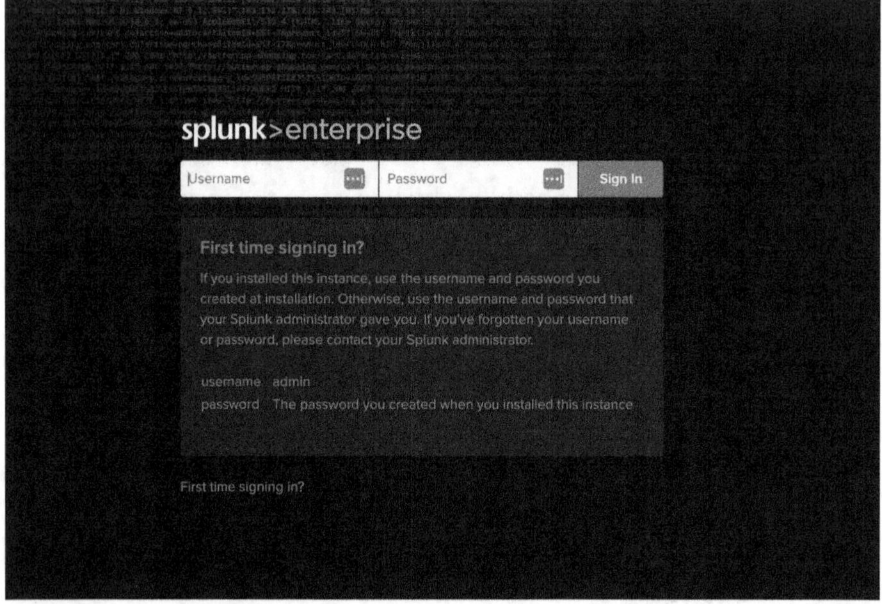

Figure 1-7. *Splunk user interface*

14. Once you are logged in, go to the Search & Reporting app, and enter the following Splunk processing command to test it.

```
index = "_internal"
```

If you get a response, you have set up the installation successfully. You should see a screen similar to Figure 1-8.

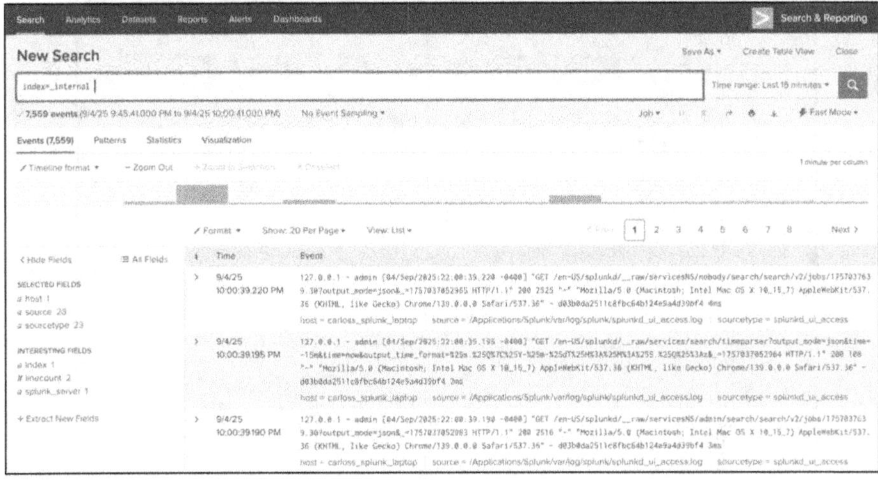

Figure 1-8. *Splunk events for index internal*

Note that host, index, linecount, source, sourcetype, splunk_server, and timestamp are a few default fields added to Splunk when indexing your data source.

This sums up the entire process of installing Splunk on macOS.

Next, let's discuss how to install it on the Windows operating system.

Installing Splunk on Windows

Windows users should follow these steps to install Splunk:

1. Sign in to your Splunk account.

2. Download the Splunk file from `www.splunk.com/en_us/download/splunk-enterprise.html`.

3. Open the downloaded file. Your screen should look similar to Figure 1-9.

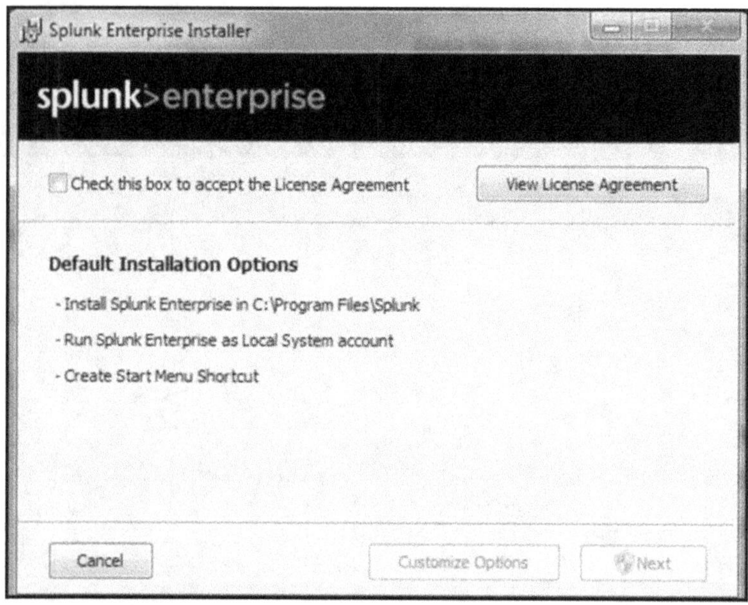

Figure 1-9. *Installing Splunk*

4. Click the check box to accept the license agreement.

5. Enter **admin** as the username, as shown in Figure 1-10.

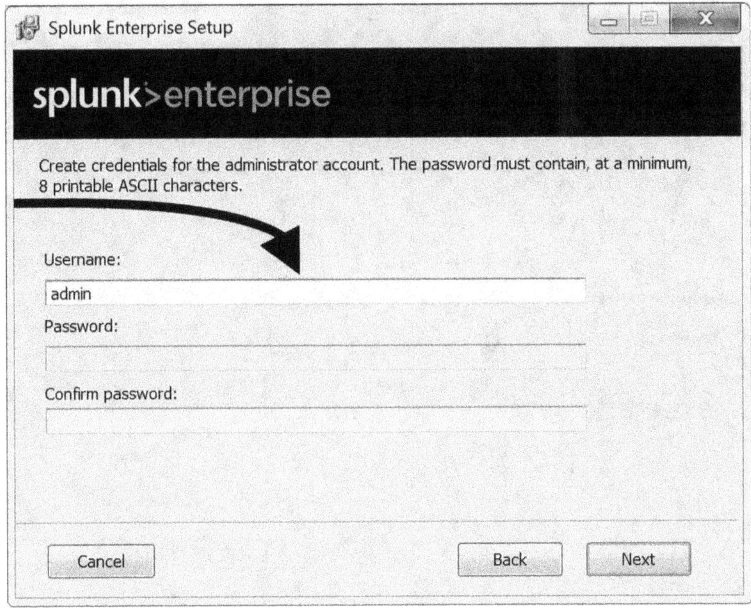

Figure 1-10. *Create administrator user*

6. Enter your password and confirm it. Then, click Next to proceed to the next step (see Figure 1-11).

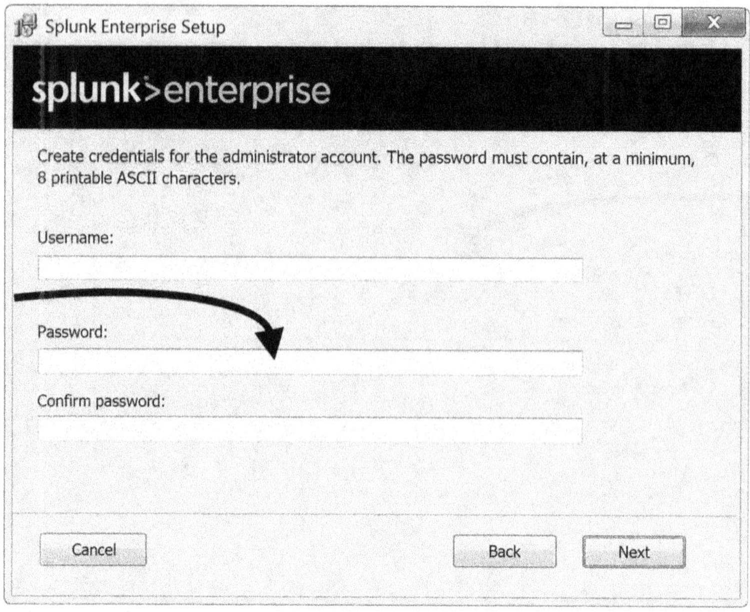

Figure 1-11. *Create password for the administrator user*

7. Click the Install button to install Splunk on your
    local machine.

8. Go to **http://localhost:8000**. In the Splunk login
    screen (see Figure 1-12), enter the username and
    password that you used in steps 5 and 6.

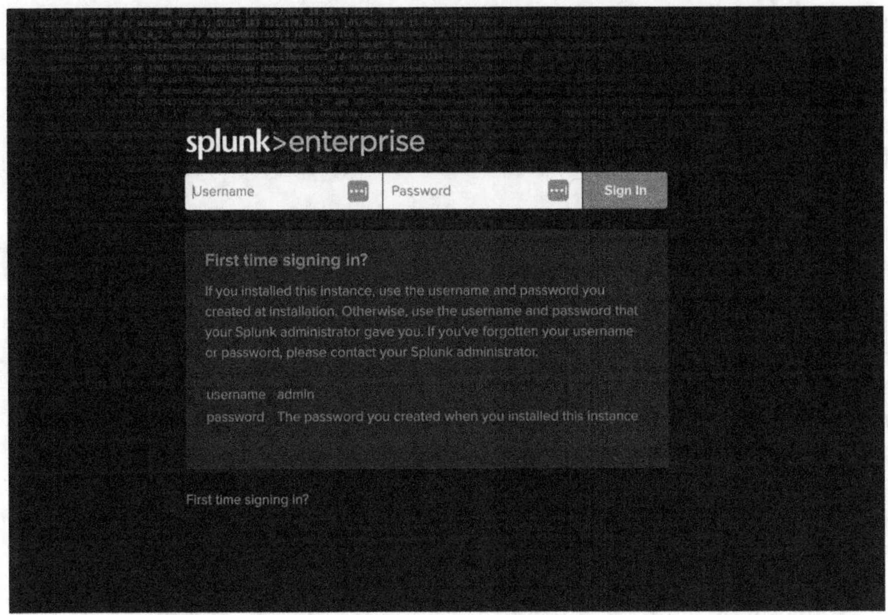

Figure 1-12. *Splunk user interface*

9. Once you are logged in, go to the Search & Reporting app and enter the following Splunk processing command to test it.

```
index = "_internal"
```

If you get a response, you have set up the installation successfully. You should see a screen similar to Figure 1-13.

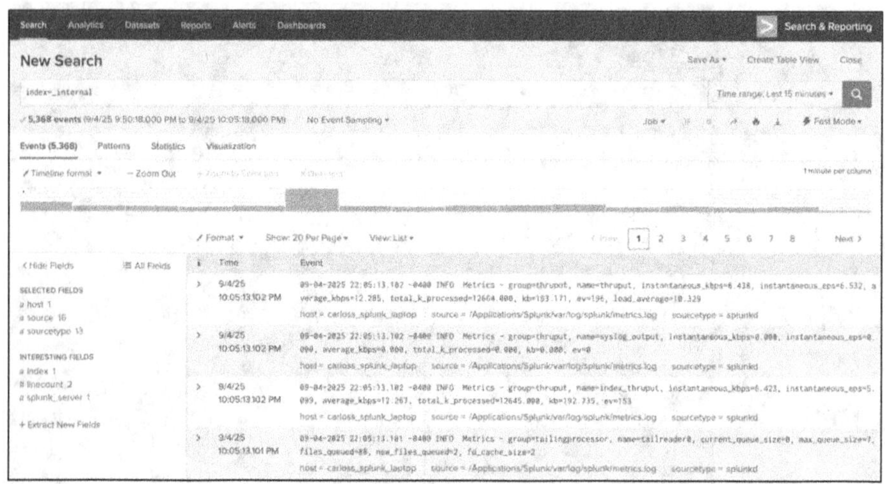

Figure 1-13. *Splunk events for index internal*

Note that host, index, linecount, source, sourcetype, splunk_server, and timestamp are a few default fields added to Splunk when indexing your data source.

With this, you have learned how to install Splunk on both macOS and Windows systems. You also learned about Splunk and its architecture and the Splunk Enterprise Certified Admin exam. In the last section of this chapter, you will learn the process of adding data to Splunk.

Adding Data in Splunk

Once Splunk is installed on your local machine, the next task is to onboard data. We can directly ingest the data by uploading the file, configuring a file monitor, or opening a TCP/UDP port. For the first time, we will upload a file and create the necessary configuration in a custom application to keep the environment clean.

1. To create a new app in Splunk, click the gear icon "Manage" next to Apps, as shown in Figure 1-14.

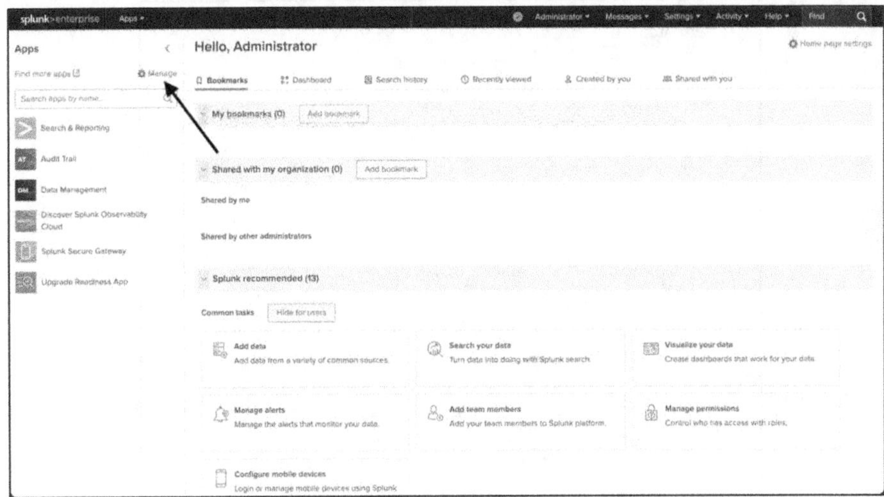

Figure 1-14. *Create a new Splunk app*

2. Click Create App. In this case, the app's name
    is *certification_book*, and the folder's name is
    also *certification_book* (see Figure 1-15). This
    folder resides in $SPLUNK_HOME/etc/apps/
    certification_book.

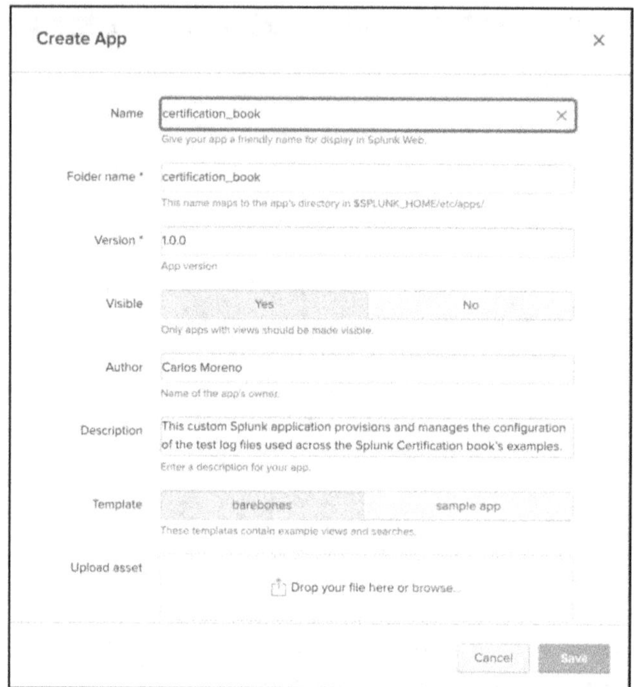

Figure 1-15. *Create certification_book app via Splunk Web*

3. As future Splunk administrators, we'll configure
    the platform to meet data-quality standards and
    optimize performance. In this step, we'll set the
    appropriate sourcetype settings in props.conf
    (timestamp extraction, line breaking, and related
    parsing options) so Splunk ingests the test logs
    accurately. Understanding and configuring the
    `props.conf` file is covered in Chapter 11.

    Once the Splunk application is created, open any
    text editor, and create a `props.conf` file in `$SPLUNK_`
    `HOME/etc/app/certification_book/local`.

```
props.conf
[client_data]
LINE_BREAKER = ([\r\n]+)\d+\s+\"\$EIT\,
SHOULD_LINEMERGE = false
TIME_PREFIX= -\s+\d{5}\s+
MAX_TIMESTAMP_LOOKAHEAD = 16
TIME_FORMAT = %m/%d/%Y %k:%M
TRUNCATE = 5000
```

4. Download the `client_data.log` file from `https://github.com/cmoreno94/splunk-certification-book/blob/main/client_data.log`.

5. Restart Splunk to make sure the changes were applied. Settings ➤ Server controls ➤ Restart Splunk.

6. Click Add Data, as shown in Figure 1-16.

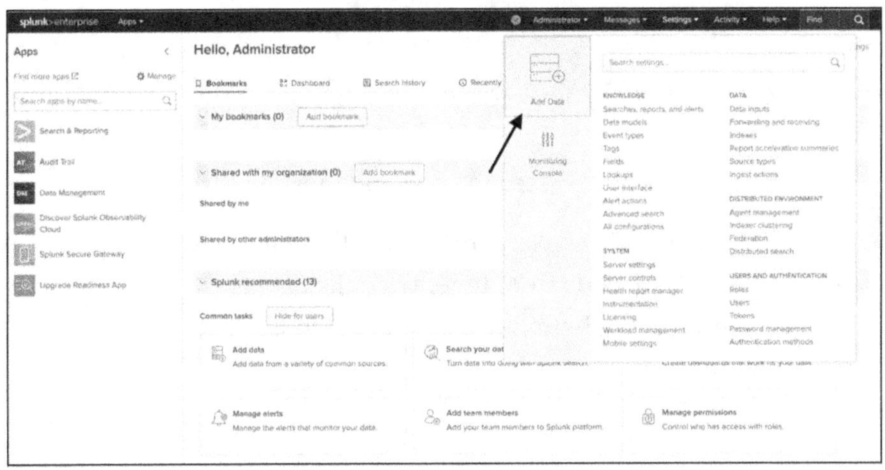

Figure 1-16. *Add data using Splunk Web*

7. Click Upload, as shown in Figure 1-17.

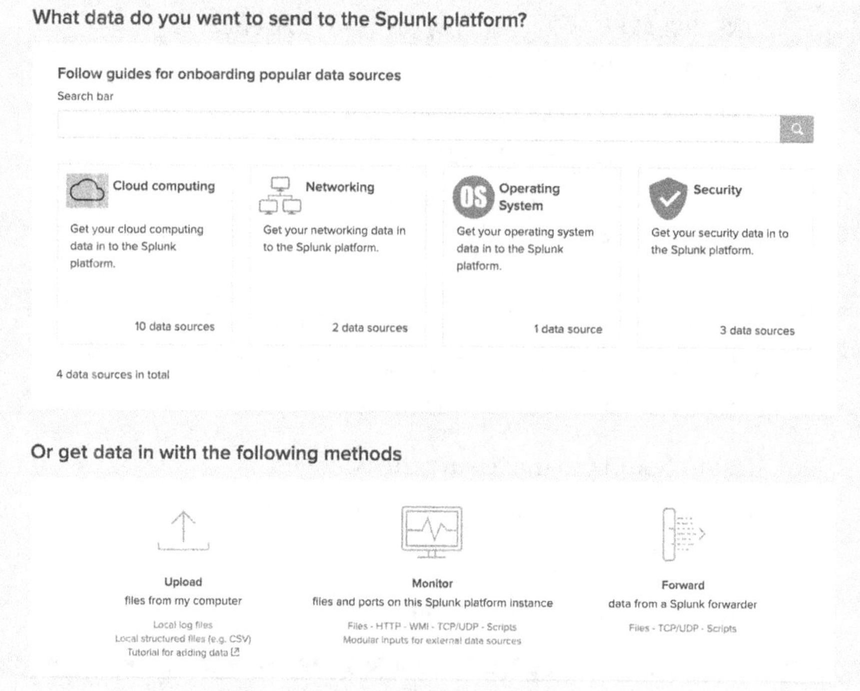

Figure 1-17. *Upload data to Splunk Web*

8. Click the `client_data.log` file, and then click Next.

9. In the Set Source Type screen, note that the current time is displayed rather than when the events occurred. For a better understanding, look at Figure 1-18.

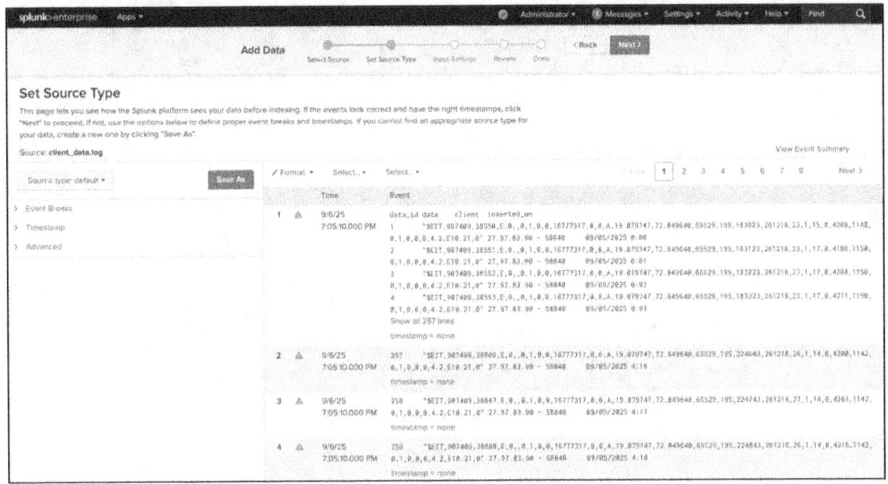

Figure 1-18. *Client data events: improper timestamp*

10. In the Source Type box, enter **client_data**, and
     select client_data as the source type. In the event
     breaks, select every line for incoming data. Now,
     the data can be transformed. The timestamp is
     selected from the Time field already present in your
     event, which is due to the `props.conf file` that you
     edited. Figure 1-19 shows the Source Type box.

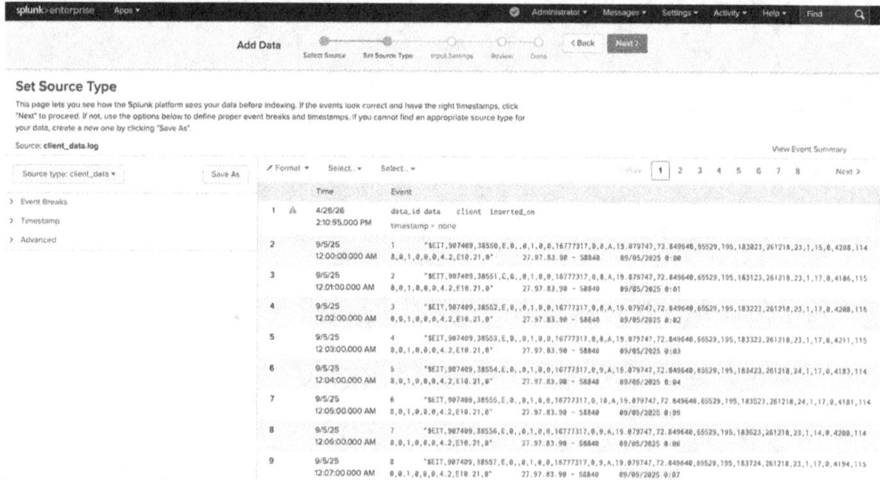

Figure 1-19. *Client data events: extracted timestamp*

11. **Header line:** In this example, the first "event" is the file header. You can exclude it by deleting the first line in the file or by dropping it at ingest with props. conf + transforms.conf. For this example, we will keep it, but in production environments, make sure only valid events are indexed.

12. Click Next. Create a new index, called transactions, in the Input settings. The transactions index stores all the events from the `client_data.log` file. Figure 1-20 shows how to create an index. You only need to enter the name.

Figure 1-20. *transactions index via Splunk Web*

13. Click the Save button after you look at all the fields.

14. Go to the certification_book app (Apps ➤ certification_book), and enter **index=transactions** in the search box. In the time bar, which is located next to the search box, select the All Time range. You see all the events in Splunk.

Congrats, you have successfully arrived at the end of Chapter 1. You now have a better idea of Splunk software and know how to input data.

Summary

This chapter highlighted the underlying concepts of Splunk by discussing its history, inception, and salient features. It provided a road map for the admin exam and explained the foundational architecture and how to install Splunk on macOS and Windows. In the last section, you saw how to play with the data in Splunk, including data onboarding, modifying source files (carrying out activities such as adding timestamps), and line breakers.

You will learn more about `.conf` files in Chapter 11. Until then, try to onboard the various types of data and play with them on Splunk.

In the next chapter, you will learn about Splunk's Search Processing Language and its various commands.

Multiple-Choice Questions

   A. The Splunk Core Certified User exam is mandatory to access the Splunk Core Certifier Power User exam.

      1. True

      2. False

   B. What is the name of the default app when you install Splunk for the first time?

      1. Home app

      2. Searching & Reporting

      3. Splunk DB app

      4. Splunkforwarder

C. What are the phases of Splunk? (Select all that apply.)

   1. Parsing

   2. Input

   3. Licensing

   4. Indexing

D. Default metadata is added to Splunk when indexing your data source.

   1. True

   2. False

E. What is the default Splunk management port?

   1. 8000

   2. 8089

   3. 8056

   4. 9997

Answers

- a. 2

- b. 1

- c. 1, 2, and 4

- d. 1

- e. 2

Further Reading

More information on how indexing works is at `https://help.splunk.com/en/splunk-enterprise/administer/manage-indexers-and-indexer-clusters/10.0/indexing-overview/how-indexing-works`.

You can look at the architecture of Splunk here: `https://docs.splunk.com/File:Architecture-new.png`.

You can read about what Splunk does with your data at `https://help.splunk.com/en/splunk-enterprise/get-started/get-data-in/10.0/introduction/how-splunk-enterprise-handles-your-data`.

CHAPTER 2

Splunk Search Processing Language

Splunk's Search Processing Language (SPL) is a user-friendly language developed to search data that has been indexed in Splunk. The language is based on Unix piping and Structured Query Language (SQL). The SPL's scope includes data searching, filtering, modification, manipulation, insertion, and deletion. You can group different events, get insight into data patterns, read real-time interactive reports, and gain business intelligence with this software. The SPL can perform various functions through a single command or a group of commands using a pipeline. Therefore, it is necessary to discuss this language before successfully operating Splunk. Chapter 3 also incorporates Splunk Search Processing Language examples.

This chapter discusses the following topics:

- The pipe operator

- Time modifiers

- Basic SPL

- Sorting results

- Filtering commands

- Reporting commands

© Carlos Moreno Buitrago, Deep Mehta 2026

C. M. Buitrago and D. Mehta, *The Splunk Core User Study Companion*, Certification Study Companion Series, https://doi.org/10.1007/979-8-8688-2501-9_2

- Filtering, modifying, and adding fields

- Grouping results

The Pipe Operator

In SPL, the pipe operator executes chains of search commands in Splunk. It is represented with the | operator. All the commands are executed from left to right in SPL, where the command to the left of the pipe operator is executed first, followed by the command to its right. Simply put, the output to the left of the pipe operator is the input to the command to the right of the pipe operator.

1. Go to the certification_book app in Splunk, and enter the following command in the search bar.

   ```
   index=transactions
   ```

2. To get the total event count in the transactions index, use index="transactions" to filter results and get all events of the transactions index. Use the pipe operator to pass the output of index="transactions" as input to get the total count of events in the transactions index. Refer to the command in the search bar.

   ```
   index=transactions | stats count
   ```

Field names in Splunk are case sensitive, but the field values are case insensitive. For example, in an index with email logs, index="email" to="deep@gmail.com" | transaction sid, the field name sid is case sensitive, but the values in sid are not case sensitive, which results in one event sid matching another event sid value.

Time Modifiers

When the indexer processes an event, the timestamp is saved as the default field _time. If you already have a timestamp in your raw data, you can update _time using `props.conf` (more about `props.conf` in Chapter 11). Whenever you execute an SPL command in Splunk, you need to be cautious about the time range. Once you fire an SPL query using a time range, only events that occurred during the time range are retrieved; the other events are filtered, so the time is the most powerful filter in Splunk, and it will help you to improve the performance of your queries. The time range picker is located to the right of the search command.

There are two major types of time modifiers: relative search and real-time search.

- A **relative search** captures data that appears in a given amount of time, such as *yesterday*. The search refreshes when the relative time boundary has passed. So a search for *yesterday* refreshes (and is, therefore, relative to) each new day.

- A **real-time search** refreshes in real time; thus, it continually drops old data from the window and acquires new data, depending on the boundaries. Each real-time search occupies a full CPU, so they are not recommended in Splunk.

Sorting results in SPL orders results in ascending order or descending order. You can choose whichever order you want based on the _time parameter.

As mentioned, time is the most efficient filter in SPL.

This section discusses the pipe operators that execute chain commands and time modifiers. In the next section, let's discuss SPL fundamentals.

Understanding Basic SPL

Understanding the basics of SPL is like learning the alphabet to best understand Splunk. SPL consists of search language syntax, boolean operators, search modes, and syntax coloring.

Search Language Syntax

Splunk has five basic search components:

- **Search terms** define the data you want to retrieve from Splunk. A term uses highlighted keywords. For example, the **index = "transactions" 58840** processing command filters all events with the number 58840 incorporated in them.

- **Commands** define what to do with the retrieved results of a search. They are used mainly for analyzing data in a result set. For example, the **index="transactions" | tail 20** processing command retrieves the last 20 results of the search; it will depend on the time picker you selected. If All Time is selected, it will return the last 20 results of the transactions index.

- **Functions** define how to chart your results. For example, the **index="_internal" | stats avg(bytes)** processing command gives the average number of bytes field in the _internal index.

- **Arguments** are the variables that you usually apply to functions to retrieve results. For example, the **index="transactions" | stats count as total_count** processing command applies the count's aggregate function as the total count.

- **Clauses** let you group results and rename field results.
  For example, the **index="_internal" | stats avg(bytes)
  as bytes by host** processing command gives the
  average bytes used in the "_internal" index used by
  each of the hosts.

Every search in Splunk is a job, and the default lifetime of that job is ten minutes; it can be extended to up to seven days. Jobs can be shared with other Splunk users in the same environment, and get the results without running the query again.

Boolean Operators in Splunk

There are three types of boolean operators in Splunk: AND, OR, and NOT. They always need to be capitalized.

The AND operator is implied between terms. Instead of writing an entire SPL statement again, you can use the boolean operator AND. For example, if you have an index called products and want to check for products that are on Amazon and Walmart, you can use the following query:

```
index="products" store="amazon" AND store="walmart"
```

The OR operator is applied to either of the terms. For instance,

```
index="products" store="amazon" OR store="walmart"
```

The NOT operator only applies to the term immediately following NOT, allowing you to omit certain results in a query. For example, we can find all the errors on a web, but skip events with the string 400 or 500.

```
index="web" error NOT (400 OR 500)
```

What about **!=** vs. the **NOT** operator? A search with != retrieves events where **fields exist**, but the field does not have the filtered value, whereas a search with the NOT operator checks if a field has a specified value and skips it, but the **field could or could not exist**.

Syntax Coloring in SPL

In Splunk, a part of the search string is automatically colored. Boolean operators, commands, arguments, and functions each have a different color; the defaults are shown in Table 2-1. Notably, you can customize or disable syntax coloring.

Table 2-1. *Syntax colors in SPL*

Argument Name	Color
Boolean operator	Orange
Commands	Blue
Argument	Green
Functions	Purple

The next section discusses sorting results.

Sorting Results

Sorting results in SPL **order results in ascending order or descending order.** You can choose whichever order you want to appear based on whatever field.

Sort

Sorting is performed using the sort command. The field name follows the sort command. By default, the sort field name sorts the results in ascending order; to sort data in descending order, you must write **sort -fieldname**. Table 2-2 describes the various types of sort commands.

Table 2-2. Sort commands

Command	Explanation
|sort field1	Sorts results in ascending order by field1.
|sort -field1	Sorts results in descending order by field1.

To limit the sorted result, you can use the limit option. The following is an example syntax query that shows how to limit the sorted result.

```
your_query | sort - field_name limit=10
```

That is all for sorting. In the following section, you will learn various Splunk filtering commands.

Filtering Commands

Filtering commands in SPL filter events based on various SPL commands, such as where, dedup, head, and tail. Let's start with the where command.

where

The where filtering command evaluates SPL to filter the results. If the result matches, then the evaluation is successful, and the result is retrieved. If the result evaluation does not match, it is unsuccessful, and the result is not retrieved. Table 2-3 explains how to use the where clause to filter results.

Table 2-3. *The where command*

Command	Explanation
lwhere field1!=value	A where statement checks if field1 is not equal to value. If so, it retrieves events that are not identical to field1.
lwhere field1=value	A where statement checks if field1 is equal to value; if it is, it retrieves the events.
lwhere field1>value	A where statement checks if field1 is greater than value; if it is, it retrieves the events.
lwhere field1<value	A where statement checks if field1 is lower than value; if so, it retrieves the events.
l where LIKE(ip, "27.%.%.%")	The where clause tries to find all the events that have an ip field range from 27.0.0.0 to 27.255.255.255.

dedup

The dedup command removes all duplicate data that falls within the same criteria. The result matches are unique to the field value provided. There won't be any other duplicate data that has an equal value to the field value.

Table 2-4 describes the various dedup commands.

Table 2-4. *The dedup command*

Command	Explanation
ldedup field1	dedup checks the field1 value for any other events with the same value; if so, it removes them and keeps the latest event.
ldedup 3 field1	dedup checks the field1 value for any other events with the same value; if so, it removes them and keeps the three latest events.

head

The head command retrieves the initial *n* events in search order. For example, head 5 retrieves the first five events.

tail

The tail command retrieves the last *n* events in search order. For example, tail 5 retrieves the last five events.

Reporting Commands

Reporting commands in SPL prepares a summary that is useful for reporting. These commands include top, rare, history, stats, untable, timechart, chart, and table.

top

The top command retrieves a table with the top value (most common) in the field values, including the field's total count and the percentage. Table 2-5 explains how to use the top command for grouping events.

Table 2-5. *The top command*

Command	Explanation
Itop field1	The top command finds the top (most common) value of all field values and retrieves the field1 table, a total count, and percentage.

By default, the output of the top command is in table format.

rare

The rare command is the opposite of the **top command** because it shows the total number of times the rare values appear in the field and their percentage of the results. Table 2-6 explains how to use the rare command for grouping events.

Table 2-6. *The rare command*

Command	Explanation
Irare field1	The rare command finds the rare value of field values and retrieves a table, total count, and the percentage of field1.

history

The history command in SPL is used to view the current user's search history.

Table 2-7 explains the history commands.

Table 2-7. *The history command*

Command	Explanation
Ihistory	Returns a table of search history
Ihistory events=true	Returns all events in the search history

table

The table command in SPL generates a table of all field names that you want to include in your report.

Table 2-8 shows the table command syntax.

Table 2-8. *The table command syntax*

Command	Explanation
Itable field1, field2	Creates a table on the only field provided

stats

The stats command calculates aggregate statistics, such as the average, count, and sum of the results. It is like SQL aggregation.

The first in line is the aggregate function.

Aggregate Functions

Aggregate functions summarize the values of each event to create a single meaningful value. Table 2-9 describes various aggregate functions in Splunk.

Table 2-9. *Aggregate functions*

Function Name	Command	Explanation
avg(field)	Istats avg(field_name)	Stats avg function returns the average number of a field.
count(field)	Istats count(field_name)	Stats count function returns a count of events for field.
distinct_ count(field)	Istats dc(field_name)	Stats dc function returns the count of the distinct values for field.
max(field)	Istats max(field_name)	Stats max function returns max value for field.
median(field)	Istats median(field_name)	Stats median function returns middle-most value for field.
min(field)	Istats min(field_name)	Stats min function returns min value for field.
mode(field)	Istats mode(field_name)	Stats mode function returns frequent value of a field.
sum(field)	Istats sum(field_name)	Stats sum function returns the sum of the field value.
var(field)	Istats var(field_name)	Stats var function returns the variance of the field value.

Event Order Functions

The event order function returns events based on the order in which the event is processed. Table 2-10 explains various event order functions.

Table 2-10. *Event order functions*

Function Name	Command	Explanation
first(field)	|stats first(field_name)	Stats first function returns first value of field seen by the stats command. The order in which the events are seen is not necessarily chronological order.
last(field)	|stats last(field_name)	Stats last function returns last value of field. The order in which the events are seen is not necessarily chronological order.

Multivalue stats and chart Functions

The multivalue stats and chart functions return the value of the field as a multivalue entry. Table 2-11 explains various multivalue stats and chart functions in Splunk.

Table 2-11. *Multivalue stats and chart functions*

Function Name	Command	Explanation
list(field)	|stats list(field_name)	Stats list function returns a list of field values, even if they are duplicated.
values(field)	|stats values(field_name)	Stats values function returns a list of distinct field values.

Time Functions

The time functions return events based on chronological order. Table 2-12 explains the various timechart functions in Splunk.

Table 2-12. *Time functions*

Function Name	Command	Explanation
earliest(field)	Istats earliest(field_name)	Stats earliest function returns first chronological value of the field.
latest(field)	Istats latest(field_name)	Stats earliest function returns last chronological value of the field.
per_day(field)	Itimechart per_day(field_name)	Timechart per_day returns the count of field per day.
per_hour(field)	Itimechart per_hour(field_name)	Timechart per_hour returns the count of field per hour.
per_minute(field)	Itimechart per_day(field_name)	Timechart per_minute returns the count of field per minute.
per_second(field)	Itimechart per_second (field_name)	Timechart per_second returns the count of field per second.

untable

The untable command converts results from a tabular format into a format similar to the statistics table.

Table 2-13 describes the untable command.

Table 2-13. *The untable command*

Command	Explanation
Iuntable fieldnameX I fieldnameY	It returns duplicate events on fieldname, but fieldvalueY has unique values from fieldnameX.

chart

The chart command is a transforming command because it returns the results in the form of a table. This command can also display data in the form of a chart. You decide what is on the X axis.

By using the chart command, you can subgroup data by using the over and the by clause.

Table 2-14 shows various chart examples and their applications.

Table 2-14. *chart examples*

Command	Explanation
Ichart count over field1	It counts the events per field1 and retrieves them in the table format. You can plot result in the graph using visualization.
Ichart count over field1 by host	It counts the events for each value in the field1 split by the value in the host field.
Ichart avg(field1) over host	It calculates the average value of field1 per each host.
Ichart avg(field1) over host by _time	It calculates the average value of field1 in the host field, split by the value in the _time field.

timechart

The timechart command is a transforming command. A timechart is a statistical aggregation command applied to the field at the Y axis to produce a chart, with time as the X axis.

Table 2-15 describes timechart commands.

Table 2-15. *timechart examples*

Command	Explanation
Itimechart count	It counts the number of events at a particular time based on the data source provided.
Itimechart count by field1	It counts the number of events at a particular time based on the field name provided.

At this point in the chapter, you are done with the pipe operators, time modifiers, the search language syntax, and the three boolean commands.

Filtering, Modifying, and Adding Fields

This section analyzes filtering, modifying, and adding fields in a report. Some of the commands are eval, rex, lookup, and fields.

eval

The eval command calculates the expression and puts the resulting value into the search field values. If the field name doesn't exist, it's created. If it already exists, its values are overwritten with the eval result. There are several functions to discuss.

Comparison and Conditional Functions

Comparison and conditional functions compare values or specify conditional statements. Table 2-16 explains the various comparison and conditional functions in Splunk.

Table 2-16. *Comparison and conditional functions*

Function Name	Command	Explanation
coalesce(<values>, ...)	| eval ip=coalesce (fieldname1,fieldname2)	This function takes one or more values and returns the first value that is not NULL.
if(<predicate>, <true_value>, <false_value>)	| eval err=if(fieldname == value, "*message_true*", "*message_false*")	It evaluates a predicate; if TRUE, return true_ value; otherwise, it returns false_value.
case(<condition>, <value>, ...)	| eval description=case (fieldname<=value, "*message*", fieldname> value AND fieldname <=value, "*message*", fieldname>value, "*message*")	It accepts different conditions and values. Returns the first value for which the condition evaluates to TRUE.
validate(<condition>, <value>, ...)	| eval n=validate(isint(port), "*message*", port >= value AND port <= value, "*message*")	It is opposite the case function.
cidrmatch(<cidr>,<ip>)	| eval isLocal=if (cidrmatch("ip_range", fieldname), "*message_true*", "*message_false*")	This function returns TRUE when an IP address, <ip>, belongs to a particular CIDR subnet, <cidr>.

(continued)

Table 2-16. (*continued*)

Function Name	Command	Explanation
like(<str>, <pattern>)	| eval is_a_foo=if(like(field, "value"), "*message_true*", "*message_false*")	It returns TRUE if text matches the pattern
match(<str>, <regex>)	| eval is_a_foo=if(match(field_value,"REGEX"), 1, 0)	It returns TRUE or FALSE based on regex matching subject.
in(<file>, <list>)	| where fieldname in("value1", "value2", "value3")	It returns TRUE, if one of the fields is validated.
null()	| eval n=null()	It takes no arguments and returns NULL.
nullif(field1, field2)	| eval n=nullif(field1,field2)	It compares the field. If field1 and field2 are the same, it returns NULL; otherwise, it returns field1.
searchmatch (<searchmatch>)	| eval test=if(searchmatch("field1=message1 AND field2=message2"), 1, 0)	It returns True if searchmatch(x) matches the event; otherwise, it returns False.

Conversion Functions

The conversion function converts numbers into strings and strings into numbers. Table 2-17 explains various conversion functions in Splunk.

Table 2-17. *Conversion functions*

Function Name	Command	Explanation
printf(<format>, <arguments>)	| eval string=printf("%04.4f %-30s",field1,field2)	It is similar to the sprintf() function in C.
tonumber(<str>, <base>)	| eval n=tonumber(fieldname)	It converts input string fieldname to a number.
tostring(<value>, <format>)	| eval time=tostring(seconds, "duration")	Converts the value in seconds to the readable time format HH:MM:SS.

Cryptographic Functions

Cryptographic functions in Splunk compute the secure hash of string values. Table 2-18 explains various cryptographic functions in Splunk.

Table 2-18. *Cryptographic functions*

Function Name	Command	Explanation
md5(X)	| eval n=md5(field)	It computes and returns the MD5 hash of X.
sha1(X)	| eval n=sha1(field)	It computes and returns the secure hash of String X based on SHA-1 hash function.
sha256(X)	| eval n=sha256(field)	It computes and returns the secure hash of String X based on SHA-256 hash function.
sha512(X)	| eval n=sha512(field)	It computes and returns the secure hash of String X based on the SHA-512 hash function.

Date and Time Functions

Date and time functions in Splunk contain functions that can calculate dates and times. Table 2-19 describes the various date and time functions in Splunk.

Table 2-19. *Date and time functions*

Function Name	Command	Explanation
now()	|eval k=now()	It takes no argument and returns time when the search is started.
relative_time(X,Y)	|eval k=relative_time (now(),"-15d@d")	It takes a UNIX time as the first argument and a relative time specifier as the second argument and returns the UNIX time.
strftime(X,Y)	eval min_ sec=strftime(_time, "%M:%S")	It takes a UNIX time as the first argument and renders the time as a string using the format specified.

Informational Functions

The informational functions contain a certain command that returns the information about values. Table 2-20 explains the informational functions in Splunk.

Table 2-20. *Informational functions*

Function Name	Command	Explanation
isint(X)	| eval n=if(isint(field),"int", "not int")	It takes one argument and returns True if the value is an integer.
isnotnull(X)	| eval n=if(isnotnull(field), "yes","no")	It takes one argument and returns True if the value is not null.
isnull(X)	| eval n=if(isnull(field), "yes","no")	It takes one argument and returns yes or no based on the argument if it's null or not null.
isnum(X)	| eval n=if(isnum(field), "yes","no")	It takes one argument and returns yes or no based on the argument value. It returns yes if the argument is a number.
isstr(X)	| eval n=if(isstr(field), "yes","no")	It takes one argument and returns yes or no based on the argument value. It returns yes if the argument is a string.
typeof(X)	| eval n=typeof(12)	It takes one argument and returns the data type of the argument.

Mathematical Functions

Mathematical functions comprise functions that can perform mathematical calculations. Table 2-21 explains the various mathematical functions in Splunk.

Table 2-21. *Mathematical functions*

Function Name	Command	Explanation
abs(X)	| eval a=abs(number)	It takes a number X, and it returns its absolute value.
ceiling(X)	| eval n=ceil(1.9)	It rounds the number X to the next highest integer.
exp(X)	| eval y=exp(3)	It returns a number and the exponential value of a function.
floor(X)	| eval n=floor(1.9)	It rounds a number X down to the nearest whole integer.
ln(X)	| eval lnBytes=ln(bytes)	It takes a number X and returns its natural logarithm.
log(X,Y)	| eval num=log(number,2)	It takes either one or two numeric arguments and returns the logarithm of the first argument X using the second argument Y as a base.
pi()	| eval area_circle=pi()*pow(radius,2)	It takes no arguments and returns the constant pi to 11 digits of precision.
sqrt(X)	|eval n=sqrt(9)	It takes one argument and returns its square root.

Multivalue eval Functions

The multivalue eval function returns multivalue fields using commands such as mvappend and mvdedup. Table 2-22 explains the types of multivalue eval functions.

Table 2-22. *Multivalue functions*

Function Name	Command	Explanation
commands(X)	|eval x=commands(search index=transactions)	It takes a search string or a field that contains a search string, such as X, and returns a multivalued field containing a list of the commands used in X.
mvappend(X,….)	| eval fullName=mvappend(initial_ values, "middle value", last_ values)	It takes N number of arguments and returns a result of all the values. It can be strings, multivalue fields, or single-value fields.
mvdedup(X)	| eval s=mvdedup(mvfield)	It takes a multivalue field and returns a multivalued field with duplicate values removed.
mvsort(X)	| eval s=mvsort(mvfield)	It takes a multivalued field and returns a field sorted lexicographically.

Statistical eval Functions

Statistical eval functions are evaluation functions to calculate statistics. Commands such as max, min, and random are types of statistical eval functions.

Table 2-23 explains these functions and their respective commands.

Table 2-23. *Statistical functions*

Function Name	Command	Explanation
max(X,Y,..)	| eval n=max(1, 3, 6, 7, "foo", field)	It takes an arbitrary input and returns the maximum result.
min(X,....)	| eval n=min(1, 3, 6, 7, "foo", field)	It takes an arbitrary input and returns the minimum result.
random()	| eval n=random()	It takes no arguments and returns a pseudo-random integer ranging from zero to $2^{31}-1$.

Text Functions

Text functions return information about strings and numeric fields in functions and nesting functions. len, lower, ltrim, and rtrim are among the types of text functions in Splunk.

Table 2-24 explains these functions and their commands.

Table 2-24. *Text functions*

Function Name	Command	Explanation
len(X)	|eval k=len(field)	It returns the field length.
lower(X)	|eval k=lower(field)	It takes one field value and returns string to lowercase.
ltrim(X,Y)	|eval k=ltrim(field,"value")	It takes one or two arguments, X and Y, and returns X with the characters in Y trimmed from the left side. If Y is not specified, spaces and tabs are removed.
rtrim(X,Y)	|eval k=rtrim(field,"value")	It takes one or two arguments, X and Y, and returns X with the characters in Y trimmed from the right side. If Y is not specified, spaces and tabs are removed.
upper(X)	|eval k=upper(field)	It takes one field value and returns string to the uppercase.

Trigonometric and Hyperbolic Functions

To calculate trigonometric and hyperbolic values, Splunk contains commands like acos, acosh, asin, and asinh. They are explained in Table 2-25.

Table 2-25. *Trigonometric and hyperbolic values*

Function Name	Command	Explanation
acos(X)	| eval n=acos(0)	It computes the arc cosine of X in the interval [0,pi] radians.
acosh(X)	| eval n=acosh(2)	It computes the arc hyperbolic cosine of X in radians.
asin(X)	| eval n=asin(1)	It computes the arc sine of X in the interval [-pi/2,+pi/2] radians.
asinh(X)	| eval n=asinh(1)	It computes the arc hyperbolic sine of X in radians.
sin(X)	| eval n=sin(1)	It computes the sine of X.
sinh(X)	| eval n=sinh(1)	It computes the hyperbolic sine of X.
tan(X)	| eval n=tan(1)	It computes the tangent of X.
tanh(X)	| eval n=tanh(1)	It computes the hyperbolic tangent of X.

Rex

The rex command extracts fields from regular expression-named groups or replaces characters in a particular field using the sed expression.

The rex command matches the value specified against the expression. Table 2-26 explains the rex command.

Table 2-26. *The rex command*

Command	Explanation
|rex field=_raw "From: <(?<from>\S+)> To: <(?<to>\S+)>"	It extracts the field values from the field selected; for example, "from" and "to " are extracted.

Mode as Sed replaces the substitute character with the given value of the field. This feature is explained as follows.

```
| rex field=fieldname mode=sed "value"
```

It replaces the current value of the field with the value you specified.

lookup

The lookup command enriches user data by adding field value combinations from tables. Lookup in SPL adds fields to event data in the results obtained from the search.

Table 2-27 explains the lookup function.

Table 2-27. *The lookup function*

Command	Explanation
| lookup lookup_name fieldname OUTPUTNEW field_lookup	The command tries to match a field's value in every event and in the lookup_name and return field_lookup from the lookup.

There are two types of lookup commands: input and output.

Input Lookup

An input lookup searches the contents of a lookup table. (More on this in Chapter 4.)

The commands of input lookup are explained in Table 2-28.

Table 2-28. *inputlookup command*

Command	Explanation
I inputlookup lookup_name	This command reads from the lookup_name lookup that is defined in transforms.conf.
I inputlookup append=t lookup_name	This command reads from the lookup_name lookup defined in transforms.conf and appends the fields to the search result.

Output Lookup

The output lookup writes fields in search results to a static lookup table file.

The output lookup commands are explained in Table 2-29.

Table 2-29. *outputlookup command*

Command	Explanation
I outputlookup lookup_name	This command writes the results of the search to lookup_name.
I outputlookup test.csv	This command writes to the test.csv lookup file under $SPLUNK_HOME/etc/system/lookups or $SPLUNK_HOME/etc/apps/<app>/lookups.

Simply put, lookup enriches the field value data through combinations.

Field

The field command keeps or removes search field results based on the conditions provided (see Table 2-30).

Table 2-30. *The field command*

Command	Explanation
| fields -field1, field2	Removes field1 and field2 from the result.
| fields field1,field2 | fields - _*	It only keeps field1 and field2 and removes the rest of the internal fields.

The last section of this chapter discusses the grouping results.

Grouping Results

Grouping results in SPL determine the total number of events in a specified _time range for a particular ID, as well as for the field that is required. It is useful to recognize the patterns from the events.

The command for grouping is the transaction command.

Transaction

The transaction command is used when you require all your events to be correlated and must define event grouping based on the start and end values. The maximum default value of a transaction is 1000. This command is resource-intensive and should be avoided where feasible.

Table 2-31 describes the various transaction commands for grouping results.

Table 2-31. *Transaction command*

Command	Explanation
Itransaction field	It displays all transactional events with the field as specified.
Itransaction field maxpause=value	It displays all transactional events with the field as specified having a maximum pause between events=value given.
Itransaction field maxspan=value	It displays all transactional events with the field as specified having maximum span between events=value given.
Itransaction field maxspan=value maxpause=value	It displays all transactional events with a field, having a maximum span between events=value and maximum pause=value given.

Summary

This chapter gave an overview of Splunk SPL. You learned about the search pipeline and the search processing language sorting command. You became familiar with the syntax coloring and result sorting features and the boolean operators AND, OR, and NOT. You also learned the filtering and reporting commands and the fields in Splunk. To learn more about these topics, please refer to `https://docs.splunk.com/Documentation/Splunk/latest/SearchReference/ListOfSearchCommands`.

The next chapter discusses field extraction, macros, and field aliases and presents an example SPL query.

Multiple-Choice Questions

A. Time is the most efficient filter in Splunk SPL.

1. True

2. False

B. What is the default timeline for any search job in Splunk?

1. Ten minutes

2. Seven minutes

3. Seven days

4. Ten days

C. By default, the output of the top command is in table format.

1. False

2. True

D. In the stats command, what do DC functions perform?

1. Returns the number of events that match the search criteria

2. Returns a count of unique values for a given field

3. Returns the sum of a numeric field

4. Lists all the values in a field

E. rex command can be used in two modes: normal and sed.

1. True

2. False

F.   Select all the functions of the stats command.

1. count

2. distinct_count

3. sum

4. chart

5. values

6. list

G.   The timechart function is a part of ___.

1. Sorting results

2. Filtering commands

3. Reporting commands

4. Grouping results

Answers

- a. 1
- b. 1
- c. 2
- d. 2
- e. 1
- f. 1, 2, 3, 5, 6
- g. 3

References

If you want to further explore Splunk searches, refer to the following reference materials:

- `https://help.splunk.com/en/splunk-enterprise/search/spl-search-reference/10.0/quick-reference/command-quick-reference`

- `https://help.splunk.com/en/splunk-enterprise/search/search-manual/10.0/specify-time-ranges/specify-time-modifiers-in-your-search`

- `https://help.splunk.com/en/splunk-enterprise/search/spl-search-reference/10.0/introduction/welcome-to-the-search-reference`

CHAPTER 3

Macros, Field Extraction, and Field Aliases

Abstract

This chapter discusses field extraction, macros, and field aliases in Splunk and explores SPL by using various queries. Field extraction in Splunk is a process that extracts fields from raw data. Splunk can extract data fields during indexing and searching. Macros in Splunk are reusable blocks (content that can be saved for future use) in which you can dynamically set the same logic for different parts or values in the dataset. Macros are useful when you want to frequently run the search command because they save you from rewriting the whole command. Field aliases in Splunk provide fields with alternate names. Field aliases normalize with various events that have similar field values.

Splunk search queries offer flexibility when working with small and big data. An SPL query helps you understand how to work with SPL and become familiar with SPL syntax.

C. M. Buitrago and D. Mehta, *The Splunk Core User Study Companion*, Certification Study Companion Series, https://doi.org/10.1007/979-8-8688-2501-9_3

This chapter covers the following topics:

- Field extraction

- Macros

- Field aliases

- An example SPL query

At the end of this chapter, you will be able to do a field extraction in Splunk using regular expressions and delimiters, create macros and field aliases, and fire a query in SPL. In other words, you will have covered up to 30% of the Splunk Core Certified Power User exam blueprint.

Let's begin with field extraction.

Field Extraction in Splunk

Field extraction is about creating new fields. Splunk extracts default fields for each event it indexes, such as host, source, and sourcetype. Field extraction extracts data fields during indexing and searching. It provides two methods to extract data: regular expressions and delimiters.

Regular Expressions

Regular expressions are mainly used for **unstructured** event data. Regular expressions are patterns used to match character combinations in strings. In Splunk Web, you just need to select the Regular Expression field and the value; it automatically generates regular expressions that match fields in other events. If the regular expression is unable to map the fields automatically, you can also manually edit the expression and discard the event accordingly.

In Chapter 1, you onboarded data in Splunk. You can use regular expressions on that data to find IPs (Internet Protocols) and ports.

The various methods for using regular expressions are explained next.

Regular Expression Using Field Extraction

In this section, you use field extraction to extract the port field from index = "transactions". The events whose values don't match the port's value will not receive the extracted field.

The following steps extract a port's value using regular expressions:

1. In Splunk Web, click Apps in the top left corner of the screen, and select the certification_book app created in Chapter 1.

2. Search events for index="transactions", and select All Time as the time range.

3. Expand the second event, click in Event Actions, and then in Extract Fields.

4. Click Regular Expression and then click Next.

5. Select the port number from the event, as shown in Figure 3-1. In the Field Name, type **port**.

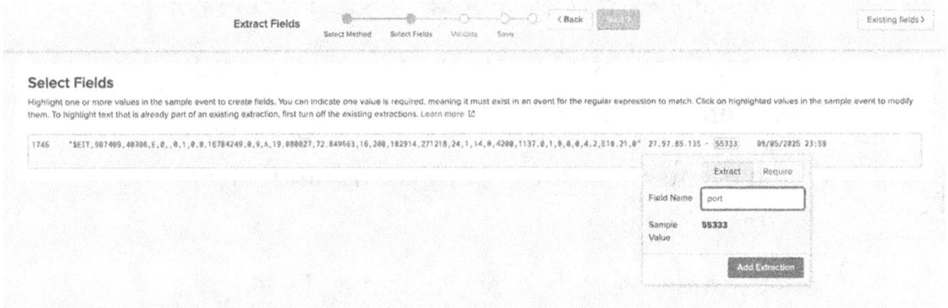

Figure 3-1. *Regular expression field extraction*

6. Click Add Extraction, and you see all events, matches, and non-matches.

7. Remember, we kept the header of the file, so this
    event won't be extracted using the current regex.
    You will see a screen similar to Figure 3-2.

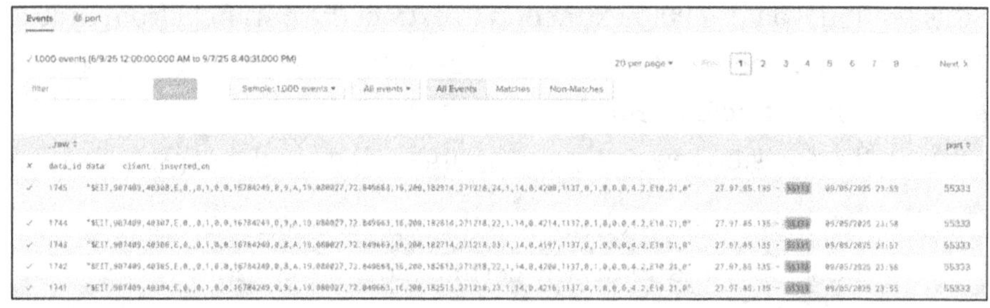

Figure 3-2. *Regular expression port extraction*

8. Click Next, Next, and then Save.

Next, let's look at inline regular expressions to extract the IP.

Inline Regular Expression Using Field Extraction

In this section, you use inline regular expressions to extract the IP field
from index = "transactions".

The following steps extract IPs in inline regular expressions:

1. In Splunk Web, click Apps in the top left corner of
    the screen, and select the certification_book app.

2. Search events for index="transactions", and select
    All Time as the time range.

3. Let's use another path to get the Extract Fields
    screen. On the left of your screen, you will see a list
    of fields (host, source, sourcetype, etc.). Go to the
    end of the list, and click Extract New Fields.

4. Select the second event, and click "I prefer to write
    the regular expression myself".

5. In the new text box, type the following regular
    expression:

    ```
    "\s+(?<ip>\d+\.\d+\.\d+\.\d+)\s
    ```

6. You should see the following results, as shown in
    Figure 3-3.

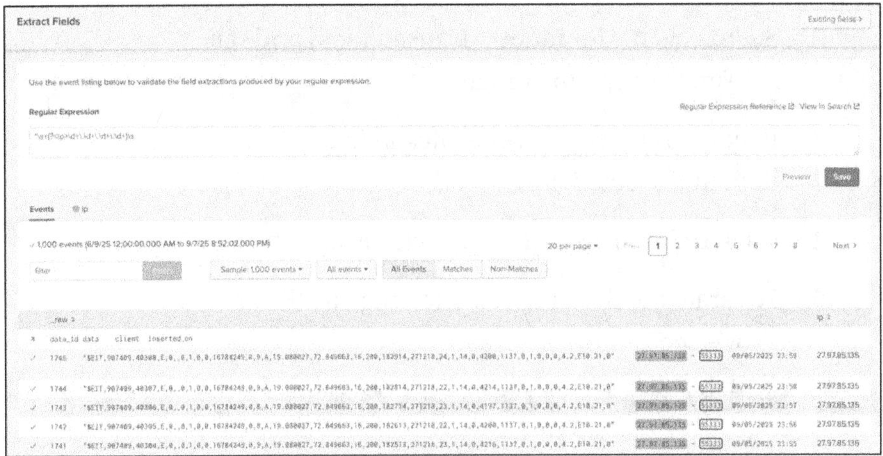

Figure 3-3. *Inline regular expressions*

7. Click Save, and then Finish.

Once the regular expression is edited, you cannot go back to Field
Extractor UI.

Next, let's discuss delimiters, another field extraction method.

Delimiters

Delimiters are used in structured data. The fields are separated by a
common delimiter, such as a comma, a space, a tab, or so forth. Delimiters
are used in CSV data and similar formats.

Delimiters Using Field Extraction

In this section, you use delimiters to extract fields from index="transactions", and use tabs to separate fields from one another. The following are the steps to use delimiters:

1. In Splunk Web, click Apps in the top left corner of the screen, and select the certification_book app.

2. Search events for index="transactions", and select All Time as the time range.

3. Expand the second event, click in Event Actions, and then in Extract Fields.

4. Click Delimiters, and then click Next.

5. Separate the delimiter by selecting Tab. Figure 3-4 shows the Delimiter screen.

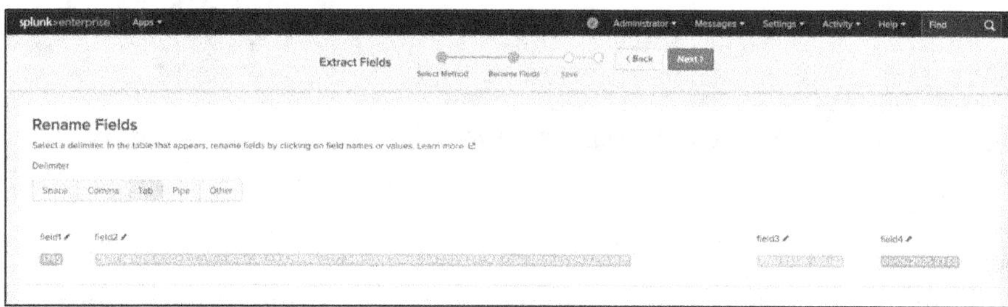

Figure 3-4. *Delimiter fields extraction*

6. Rename all fields as follows:

    - field1=id

    - field2=lat_long

    - field3=ip_port

    - field4=timestamp

Figure 3-5 shows the renaming process.

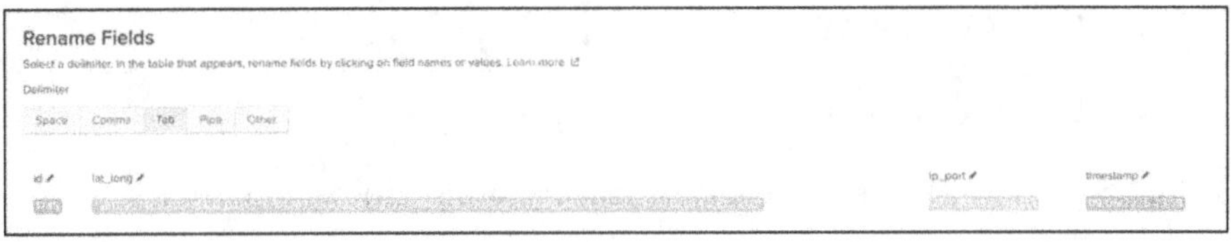

Figure 3-5. *Renaming fields in delimiters*

7. Click Next, and save Report name as get_client_
    data_fields.

8. Click Finish.

Use delimiters when your event contains structured data, such as CSV files, and so forth.

Pro tip   We have created Knowledge Objects (KOs), and depending on the level of permissions you set (by default, owner), you can go to `$SPLUNK_HOME/etc/users/admin/certification_book/local/props.conf` or `$SPLUNK_HOME/etc/apps/certification_book/local/props.conf` and check the KOs directly in the configuration files. You can edit those files and create/update/remove multiple KOs quickly.

Let's discuss search command macros in the next section.

Macros

Macros in Splunk can be a full search command or part of the search command in SPL. Macros are useful when you want to frequently run the search command because you do not need to rewrite the whole search.

There are two ways to create macros in Splunk:

- Using Splunk Web

- Using a .conf file

Macros are reusable knowledge objects since multiple people and multiple apps can use them.

Create a Macro Using Splunk Web

Use Splunk Web to create a macro for the certification_book app, and make a macro for a transaction in which the max pause is 20 minutes.

To create macros using Splunk Web, follow these steps:

1. Click Settings and go to Advanced Search.

2. Select Search Macros and click New Macro.

3. In Destination app, type certification_book.

4. In Name, type **session**.

5. In Definition, type **transaction maxpause=20m ip**.

6. Click Save.

   Figure 3-6 shows the Add New page where these steps are performed.

Figure 3-6. *Create a macro*

7. To test the macro, go to the search bar, and type the following command:

```
index="transactions" | `session` | stats sum(eventcount)
as session
```

Figure 3-7 shows the output.

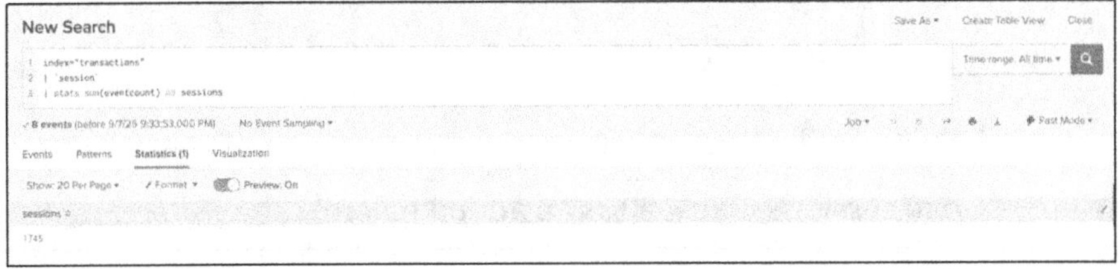

Figure 3-7. *Search output*

Note   To run macros in SPL, you must use ` macro_name`.

Next, let's look at another method for creating macros.

Create a Macro Using the .conf File

To create a macro using the `.conf` file in Splunk, go to `$SPLUNK_HOME/etc/apps/<app_name>/local/macros.conf`. If it is not there, create it; if it is there, edit it.

In this section, you create a macro for the certification_book app using the `.conf` file. The macro would be a transaction with a max pause of 20 minutes. Go to `$SPLUNK_HOME/etc/apps/certification_book/local/`. If `macros.conf` is there, edit it; otherwise, you need to create it.

Once the `macros.conf` file is ready, follow these steps:

1. Add the following block:

    ```
    [sessions_pause20]
    definition = index="transactions" | transaction
    maxpause=20m ip | stats sum(eventcount) as sessions
    ```

2. Restart Splunk.

3. Go to Settings.

4. Go to Advanced Search.

5. Search macros.

6. Go to sessions_pause20 as an added macro using .conf.

    The working Search Macros page is illustrated in Figure 3-8.

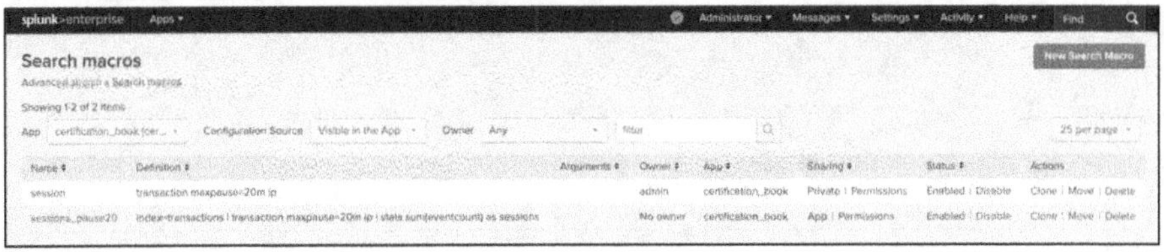

Figure 3-8. *Illustrated macros*

7. To test this macro, go to the search bar, and type the
    following command:

    `` `sessions_pause20` ``

Figure 3-9 shows the output.

Figure 3-9. *Search results for sessions_pause20 macro*

Macros can contain variables as well, which can be passed once we
use the macro. In Figures 3-10 and 3-11, you can see how to create a macro
with one argument and how to use it.

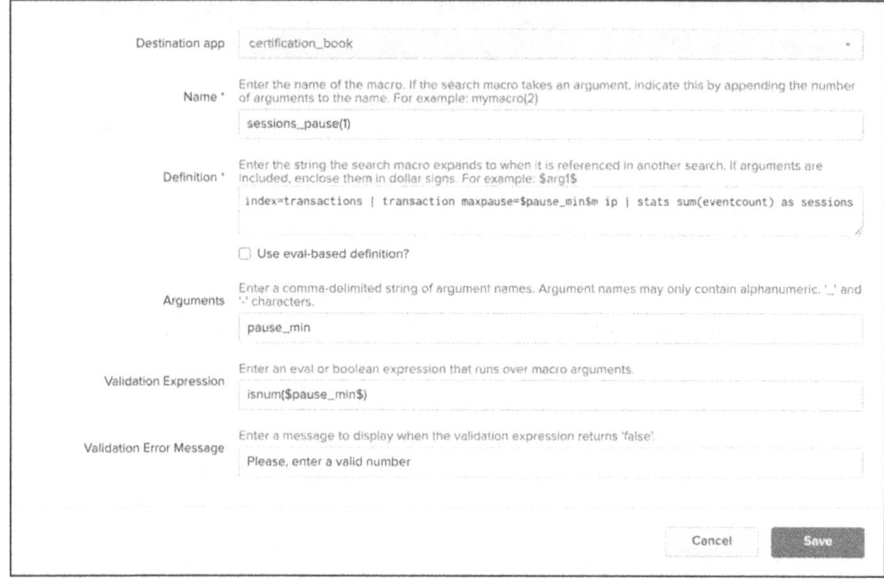

Figure 3-10. *Macro sessions_pause() with one argument*

Figure 3-11. *Search results for sessions_pause(5) macro*

Note   To expand a search macro and view its full definition within the search bar, you can utilize a keyboard shortcut:

Mac OSX: Command-Shift-E

Linux or Windows: Control-Shift-E

Next, let's discuss field aliases.

Field Aliases in Splunk

A field alias in Splunk is an alternate name assigned to an existing field, allowing users to search for events containing that field using the alias instead of the original name. This feature is useful for normalizing field names across different data sources or simplifying complex field names for easier searching. Let's look at setting up field aliases in Splunk environments.

Setting Up Field Aliases

In this section, you upload two data sources: transactions_log.txt and transactions_advertise.csv. We're using two new file types (TXT and CSV), and Splunk will ingest them without special handling. In transactions_log.txt, you have the "unit_id" field, and in transactions_advertise.csv, you have the "id" field. You create unit_id and id field aliases.

You can download transactions_log.txt from `https://github.com/cmoreno94/splunk-certification-book/blob/main/transactions_log.txt`.

To set up the field aliases, follow these steps:

1. Go to Settings ➤ Add Data ➤ Upload, and select transactions_log.txt.

2. Splunk can automatically detect the timestamp and event breakers, but it will decrease the performance of your instances. The best practices are to configure them for each sourcetype, improving the performance of your Splunk instance. Since it is not a production environment, we can keep the Source Type as default and Event Breaks as Every line.

Figure 3-12 shows the Set Source Type page.

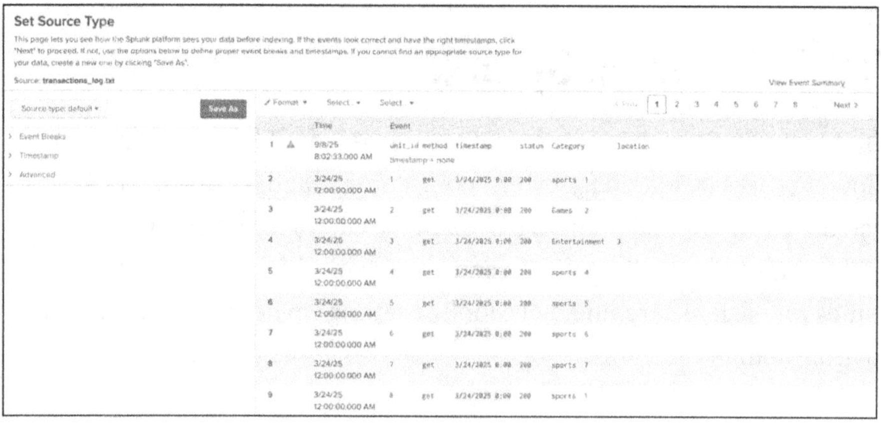

Figure 3-12. transactions_log events

3. Click Next, and create the new sourcetype=transaction_data.

4. In the Inputs Settings screen, select index=transactions.

5. Click in Review, and then in Submit.

You can download transactions_advertise.csv from `https://github.com/cmoreno94/splunk-certification-book/blob/main/transactions_advertise.csv`.

1. Go to Settings ➤ Add Data ➤ Upload, and select transactions_advertise.csv file.

2. Set Source Type as csv. The Set Source Type page is shown in Figure 3-13.

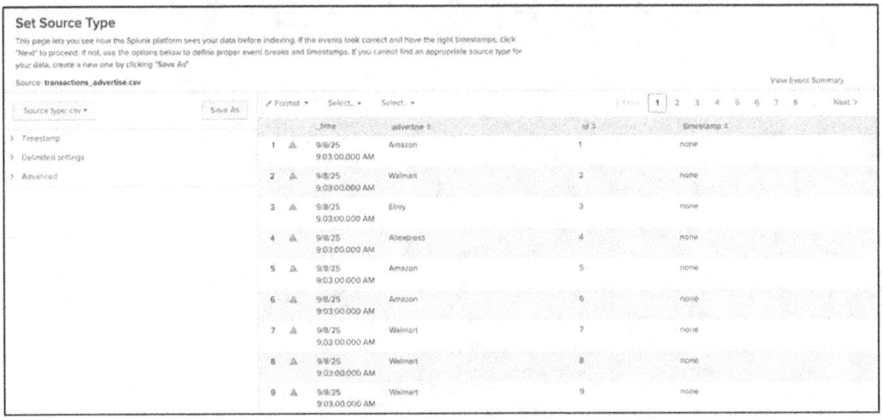

Set Source Type

This page lets you see how the Splunk platform sees your data before indexing. If the events look correct and have the right timestamps, click "Next" to proceed. If not, use the options below to define proper event breaks and timestamps. If you cannot find an appropriate source type for your data, create a new one by clicking "Save As".

Source: transactions_advertise.csv

		_time	advertise ¢	id ¢	timestamp ¢
1		9/8/25 9:03:00.000 AM	Amazon	1	none
2		9/8/25 9:03:00.000 AM	Walmart	2	none
3		9/8/25 9:03:00.000 AM	Ebuy	3	none
4		9/8/25 9:03:00.000 AM	Aliexpress	4	none
5		9/8/25 9:03:00.000 AM	Amazon	5	none
6		9/8/25 9:03:00.000 AM	Amazon	6	none
7		9/8/25 9:03:00.000 AM	Walmart	7	none
8		9/8/25 9:03:00.000 AM	Walmart	8	none
9		9/8/25 9:03:00.000 AM	Walmart	9	none

Figure 3-13. *transactions_advertise.csv events*

3. Click Next, select index=transactions.

4. Click in Review and Submit.

For transactions_log.txt field extraction, do the following:

1. Navigate to the certification_book app:

   ```
   index=transactions sourcetype=transaction_data
   ```

2. As we discussed in Chapter 2, use delimiters to extract field with delimiter=Tab, and then provide the following field names. Expected results are shown in Figure 3-14.

   ```
   field1=unit_id
   field2=method
   field3=timestamp
   field4=status
   field5=category
   field6=location
   ```

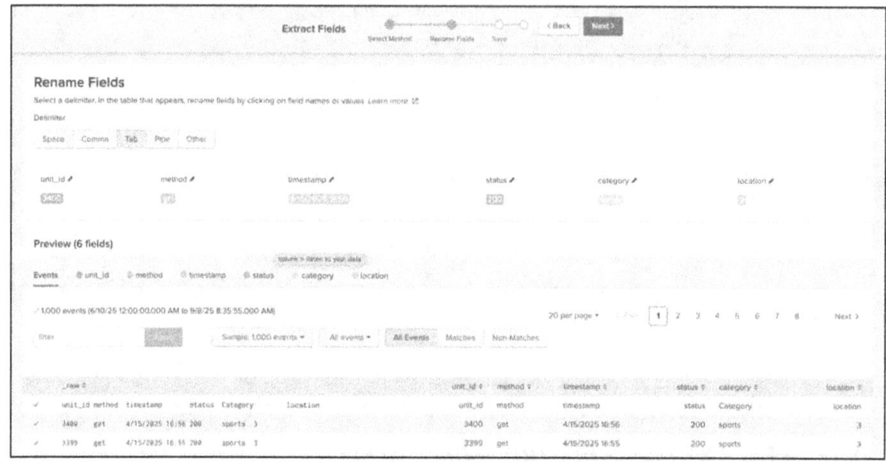

Figure 3-14. *Field extractions for transaction_data sourcetype*

3. Save the report as get_transactions_data_fields.
    Click Finish.

For transactions_advertise.csv field extraction, we do not need
to create the field extractions since the sourcetype is csv. Splunk will
automatically extract the field names from the header. You can check it by
running the following query:

```
index=transactions sourcetype=csv | table id advertise
```

In transactions_log.txt, "unit_id" is a common field, and in
transactions_advertise.csv, "id" is a common field, so you can use field
aliases to standardize the field names and correlate events.

Follow these steps to create a field alias for both fields and get a new
field called transaction_id:

1. Go to Settings ➤ Fields.

2. Go to Field aliases ➤ Add New.

3. In Destination app, type certification_book.

4.   In Name, type **transaction_id**.

5.   In Apply to, select sourcetype.

6.   In named, type transaction_data.

7.   In Field aliases, type **unit_id=transaction_id**.

The Add New screen is shown in Figure 3-15.

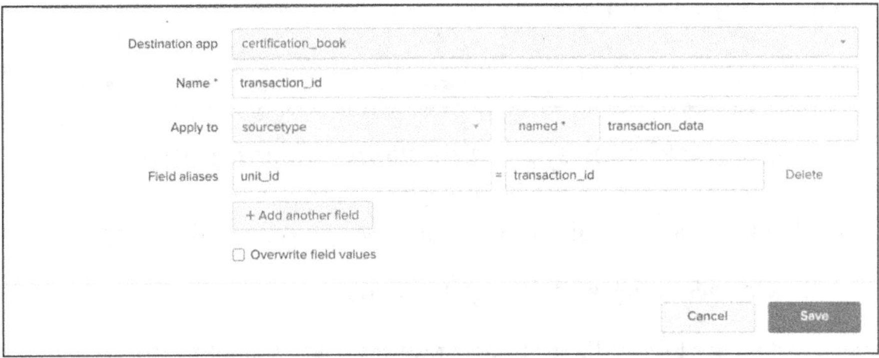

Figure 3-15.   *Field aliases for transaction_data sourcetype*

Now, the FIELDALIAS-transaction_id will be applied to all the events with sourcetype=transaction_data. In the next example, we will apply the field alias at the source level, since sourcetype=csv can be used in other data sources.

Like the prior example, add another field alias having similar field values.

1.   In Destination app, type certification_book.

2.   In Name, type **transaction_id_advertise**.

3.   In Apply to, select Source, and in named, type transactions_advertise.csv.

4.   In Field aliases, type **id=transaction_id**.

The working screen is shown in Figure 3-16.

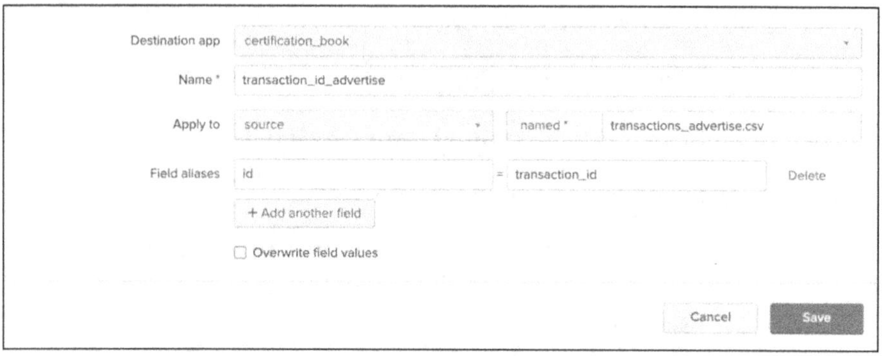

Figure 3-16. *Field aliases for transactions_advertise.csv source*

To check the success of the commands, refer to the following steps:

1. Go to Settings and then go to Fields.

2. On the Field Aliases page, you find that two field aliases have been added, as shown in Figure 3-17.

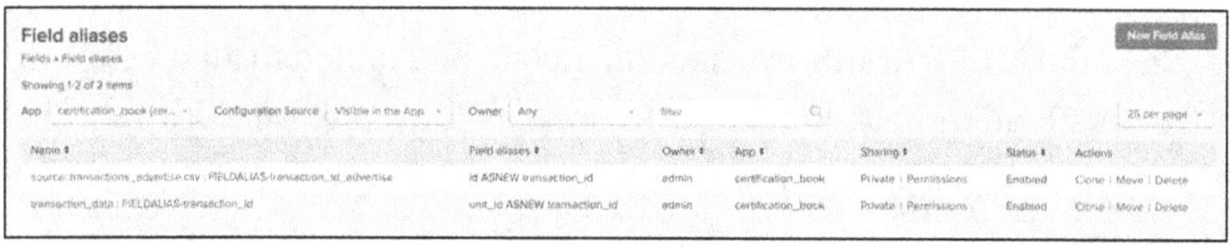

Figure 3-17. *Field aliases created*

To test the field aliases that you added, type the following SPL command in the search bar and Time range=All time:

```
index=transactions transaction_id=* | stats count by sourcetype
```

The page to test the field aliases is shown in Figure 3-18.

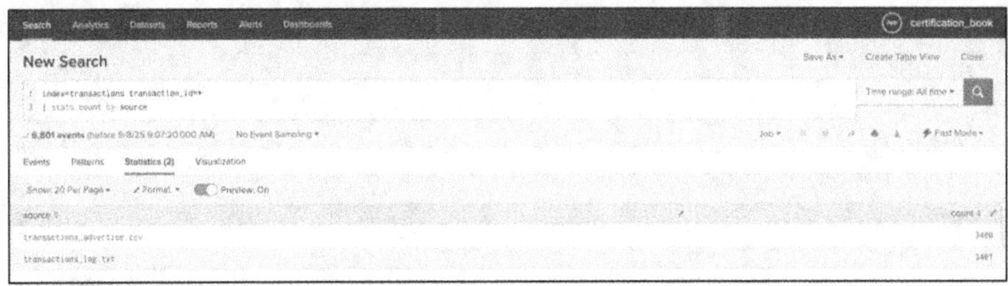

Figure 3-18. *Field alias for transaction_id in search*

Field aliases provide normalization over any field (host, source, sourcetype, etc.).

Next, let's walk through a Splunk search query that helps correlate data.

Splunk Search Query

Splunk search queries correlate data and are used in data searching, filtering, modification, manipulation, enrichment, insertion, and deletion. They provide flexibility when using small and big data. You can also create, analyze, and visualize charts in Splunk.

1. Try to write a query in which the user wants to know the events related to the transaction ID 100. This is the solution.

    ```
    index=transactions transaction_id=100
    ```

    The solution is shown in Figure 3-19.

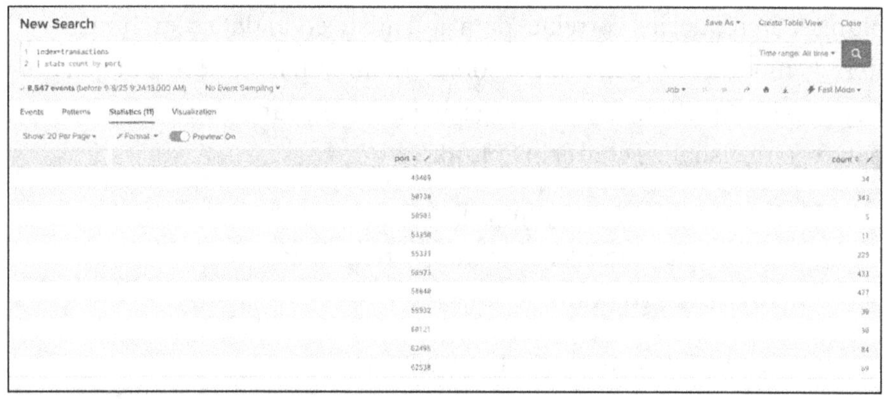

Figure 3-19. *Results of example 1*

2. Write a Splunk query to get the event count from each port.

    This is the solution.

    ```
    index=transactions |stats count by port
    ```

    The solution is shown in Figure 3-20.

Figure 3-20. *Results of example 2*

3. Write a query to get the number of distinct advertise values.

    ```
    index=transactions sourcetype=csv | stats
    dc(advertise) as dc_advertise
    ```

The solution is shown in Figure 3-21.

Figure 3-21. *Results of example 3*

4. Pro example: Get the total transactions per advertise, when the transaction was successful (status=200).

   This is the solution.

   ```
   index=transactions sourcetype IN (csv,
   transaction_data)
   | fields transaction_id status advertise
   | stats values(status) as status values(advertise) as
     advertise by transaction_id
   | where status=200
   | stats count by advertise
   ```

   The solution is shown in Figure 3-22.

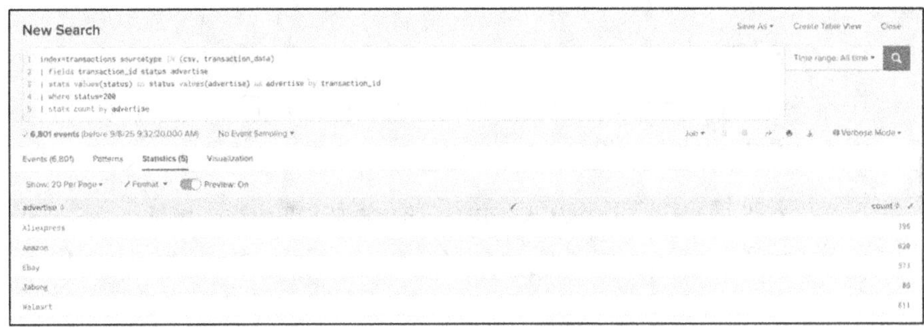

Figure 3-22. *Results of example 4*

Pat yourself on the back because you have successfully learned how to create macros, field extractions in Splunk using regular expressions and delimiters, field alias, and completed four great SPL queries.

Summary

This chapter covered field extraction in Splunk to extract data fields during indexing and searching. You learned that macros are useful when you need to frequently run the search command because they prevent rewriting the entire command. You also learned about field aliases, which normalize various events with similar field values. Also, a Splunk search query provides flexibility when working with small and big data.

According to the Splunk Core Certified Power User exam blueprint, 10% is on field extraction, 10% is on macros, and 10% is on field aliases.

In the next chapter, you will learn about Splunk tags and lookups and create alerts.

Multiple-Choice Test Questions

    A. Macros can be created only via Splunk Web.

        1. True

        2. False

    B. Which directory contains the configuration files in Splunk?

        1. SPLUNK_HOME/etc/apps/<app>/default/

        2. SPLUNK_HOME/var/log/

        3. SPLUNK_HOME/secret/

        4. SPLUNK_HOME/etc/passwd

C. Once a regular expression is edited manually, can you go back to the Field Extractor UI?

1. True

2. False

D. Are delimiters mostly used in structured data?

1. True

2. False

E. Field aliases normalize data over which default fields? (Select all that apply.)

1. Host

2. Source

3. Source type

4. Events

F. Fields appear in the Interesting Fields lists, if they appear in at least _____ of events.

1. 10%

2. 80%

3. 40%

4. 20%

Answers

- A: 2

- B: 1

- C: 2

- D: 1

- E: 1, 2, and 3

- F: 4

References

- *Splunk Operational Intelligence Cookbook* by Josh Diakun, Paul Johnson, Derek Mock (Packt Publishing, 2018).

- https://docs.splunk.com/Documentation/UnixApp/latest/User/Searchmacros

- https://help.splunk.com/en/splunk-enterprise/manage-knowledge-objects/knowledge-management-manual/10.0/search-macros/define-search-macros-in-settings

- https://help.splunk.com/en/splunk-enterprise/manage-knowledge-objects/knowledge-management-manual/10.0/use-the-settings-pages-for-field-extractions-in-splunk-web/use-the-field-extractions-page

CHAPTER 4

Tags, Lookups, and Correlating Events

In the previous chapter, you learned how to extract fields from Splunk using delimiters and regular expressions to create macros and field aliases. You also performed a few Search Processing Language commands to improve your SPL skills. In this chapter, you deal with Splunk tags and lookups and create various reports and alerts. Splunk tags are pairs of nomenclature-added values that assign names to a specific field and its value combination. Splunk lookups enhance data by adding a field/value combination from any internal or external data source. Reports are saved searches or a pivot, which represent related statistics and visuals for later reuse. Meanwhile, alerts are triggered when a search result satisfies any condition.

The following are the topics in this chapter:

- Splunk lookups

- Splunk tags

- Creating reports in Splunk

- Creating alerts in Splunk

By the end of this chapter, you'll have covered at least 10% of the Splunk Core Certified Power User exam blueprint and 17% of the Splunk Core Certified User exam blueprint.

© Carlos Moreno Buitrago, Deep Mehta 2026
C. M. Buitrago and D. Mehta, *The Splunk Core User Study Companion*, Certification Study Companion Series, https://doi.org/10.1007/979-8-8688-2501-9_4

Splunk Lookups

Lookups are search-time knowledge objects that enrich your events by adding fields from an internal or external dataset. Given one or more **key fields** in the event (e.g., `error_code=404` or `src_ip=8.8.8.8`), Splunk consults a lookup table and writes back additional attributes (e.g., `error_text=" Not Found"` or `provider="Google"`). You can apply a lookup **ad hoc** in a search with the lookup command or configure an automatic lookup so the enrichment happens for every matching event without modifying your SPL.

There are four types of lookups in Splunk, as represented in Figure 4-1.

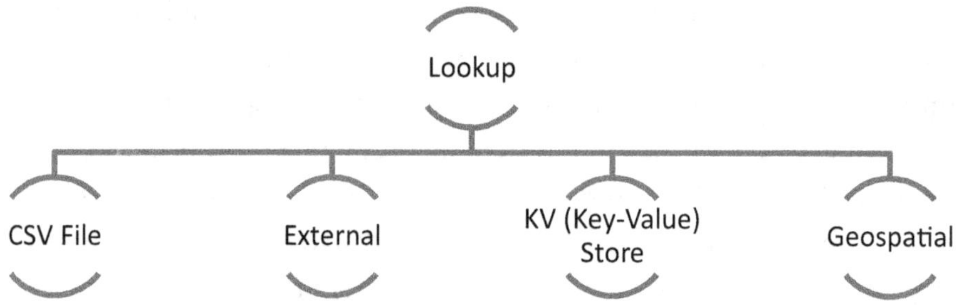

Figure 4-1. *Splunk lookups*

Each of these lookup types is explained as follows:

- **CSV (file-based) lookups:** A simple table to pull key/value pairs from CSV files. Best for small/medium, relatively static datasets. Therefore, they are also called **static lookups**. Each column in a CSV table is interpreted as a potential field value.

- **KV Store lookups:** It is a schemaless collection managed by Splunk KV Store. Similar to a MongoDB database. Good for larger or frequently changing data, upserts, and app-managed content. For example, the app's checkpoints are managed in a KV store.

- **Geospatial lookups:** Maps coordinates or other keys to geographic shapes/regions using KMZ/KML datasets. Create queries that return results, which Splunk platform can use to generate a choropleth map visualization.

- **External lookups** populate event data from an external source. They can be **Python scripts** or **binary executables** that get field values from an external source. They are also called **scripted lookups**.

There are three different lookup knowledge objects in Splunk:

- Lookup table files

- Lookup definitions

- Automatic lookups

Field values are case sensitive by default.

Lookup Table Files

You use the lookup table file in Splunk when you need to define a lookup using a CSV file or a GZ file, depending on their availability. In this section, you define a sample lookup table, which is called location.csv. You can define your files in a preexisting lookup table file.

1. Download location.csv from `https://github.com/ cmoreno94/splunk-certification-book/blob/ main/location.csv`. location.csv has two columns, location and city as shown in Figure 4-2.

```
location,City
1,New york
2,San Francisco
3,Chicago
4,Washington D.C.
5,Seattle
6,Boston
```

Figure 4-2. *location.csv*

2.   Go to Settings on Splunk Web, and then go to
     Lookups.

3.   Go to Lookup table files and click Add New.

4.   In Destination app, select the certification_
     book app.

5.   Upload the location.csv file. For your reference, a
     sample image is shown in Figure 4-3.

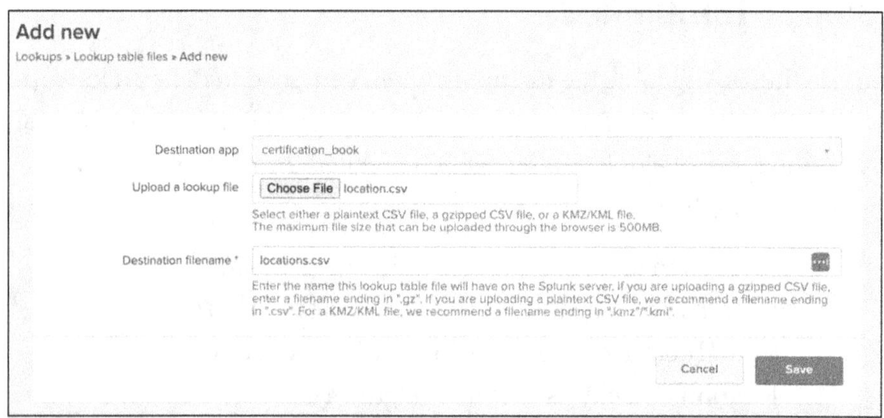

Figure 4-3. *Create the new lookup table*

6. Once the lookup is created, go to Lookup Table Filter, where you find location.csv.

7. Click Permission on the location.csv file. You have the option to edit various roles. Select the **All apps (system)** option. (Chapter 6 covers knowledge object permissions.)

8. Run the following query to validate the new lookup table, returning seven results:

```
| inputlookup locations.csv
```

Lookup Definitions

A lookup definition provides a lookup name and a path to find the lookup table. Lookup definitions can include extra settings such as matching rules or restrictions on the fields that the lookup is allowed to match. One lookup table can have multiple lookup definitions. In this section, you create a lookup named *cities* for the location.csv lookup file defined in the lookup table file.

Follow these steps to create lookup definitions:

1. Go to Settings and select Lookups.

2. Go to Lookup definitions and click Add New.

3. In Destination app, select the certification_book app. Name it **cities**. In Type, select File-based. In Lookup file, select location.csv. For a better understanding, refer to Figure 4-4.

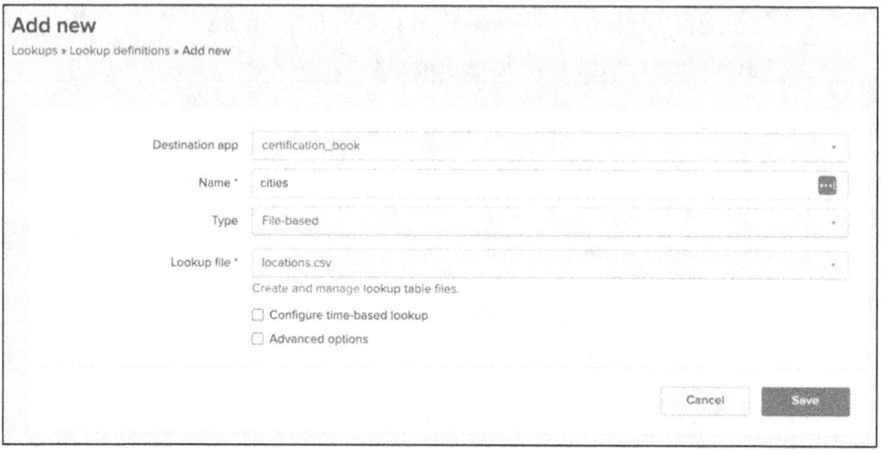

Figure 4-4. *Lookup definitions*

4.   Click the Save button, and change the countries
     file's permissions to All apps.

5.   Run the following query to validate the new lookup
     table, returning seven results:

```
| inputlookup cities
```

In the next section, you map the lookup input field with the lookup
output field.

Automatic Lookups

Automatic lookups apply a lookup to all searches at search time. The
automatic lookup requires a lookup definition to be configured, and it
will map the lookup definition fields with the fields in the search. In this
section, you depict the field location events from the transactions_log.txt
source type with the city column in the location.csv.

The following steps create an automatic lookup:

1.   Go to Settings and click Lookups.

2. Select Automatic Lookups and then select Add New.

3. In Destination app, select certification_book. In
    Name, enter **match_locations**. In Lookup table,
    select the countries lookup definition that you
    created in the previous step. In Apply to, select
    sourcetype named as transaction_data. In Lookup
    input fields, type **location = location**. In Lookup
    output fields, enter **City = city**. Refer to Figure 4-5.

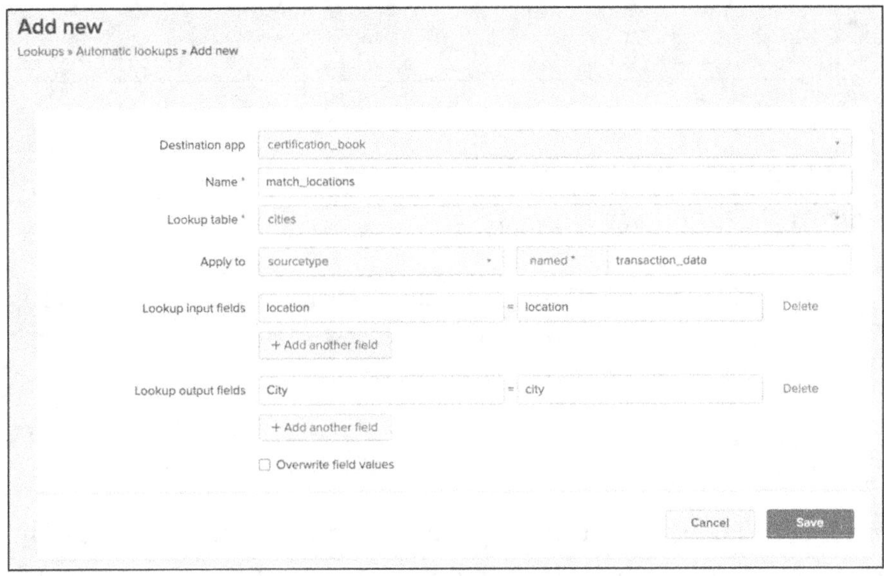

Figure 4-5. *Splunk automatic lookup*

4. Update the permissions to All apps, and run the
    following search in the certification_book app:

```
index=transactions sourcetype=transaction_data
earliest=0
| table _time location city
```

Keep in mind the "city" field is not in transactions_log.txt, but the automatic lookup matches the location field between the transactions_log.txt events and the cities lookup definition and automatically returns the city field to be consumed in the search. You can find cities in a field with events mapped (see Figure 4-6).

_time ^	location ⇕	city ⇕
2025-03-24	6	Boston
2025-03-24	1	New york
2025-03-24	6	Boston
2025-03-24	6	Boston
2025-03-24	3	Chicago
2025-03-24	6	Boston
2025-03-24	5	Seattle

Figure 4-6. Table with city field

In Splunk, admins can change the case_sensitive_match option to false in the lookup definition and/or transforms.conf.

Splunk Tags

Tags are also knowledge objects in Splunk. They allow the **user to search for an object with a particular combination of conditions**, such as a field value. Tags make data more understandable. More than one tag can be created for a specific field/value combination.

Create Tags in Splunk Using Splunk Web

The Splunk Web interface conveniently creates tags. In this section, you create a tag named *privileged* for location=3 in the transactions_log.txt source. Follow these steps to create tags:

1. Go to the certification_book app. In the search bar,
    type the following command:

    ```
    index=transactions sourcetype=transaction_data
    location=3 earliest=0
    ```

2. Explore an event and go to the field named location.

3. In Actions, click the down arrow and select
    Edit Tags.

4. In Tag(s), type **privileged**. The Field Value is
    location=3, as shown in Figure 4-7.

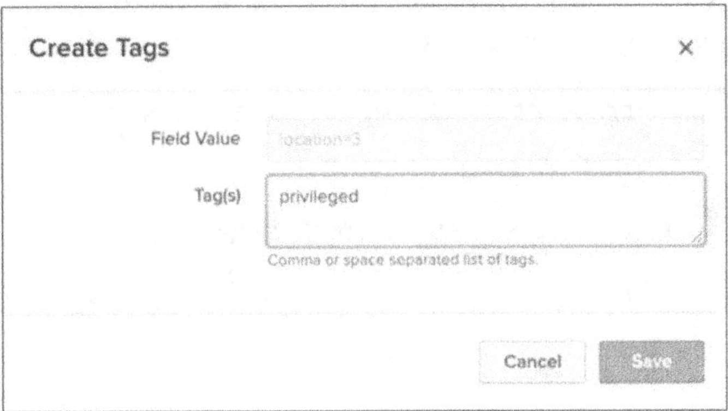

Figure 4-7. *Privileged tag*

5. You have successfully created a tag. You can check it
    with the following search command:

    ```
    index=* tag=privileged earliest=0
    ```

Tag Event Types in Splunk Web

Tag event types in Splunk add extra information to events. In this section, tag event type named *privileged* is located in the transactions_log. txt source.

Follow these steps to create this:

1. Go to Settings and then select Event types.

2. Create a new event type.

3. In Destination App, select certification_book. In Name, enter **vip**. In Tag(s), enter **vip**. In Color, select yellow. In Priority, select 1. In Search string, type the search query as follows:

```
index=transactions sourcetype=transaction_data
location=6
```

Figure 4-8 represents the working page.

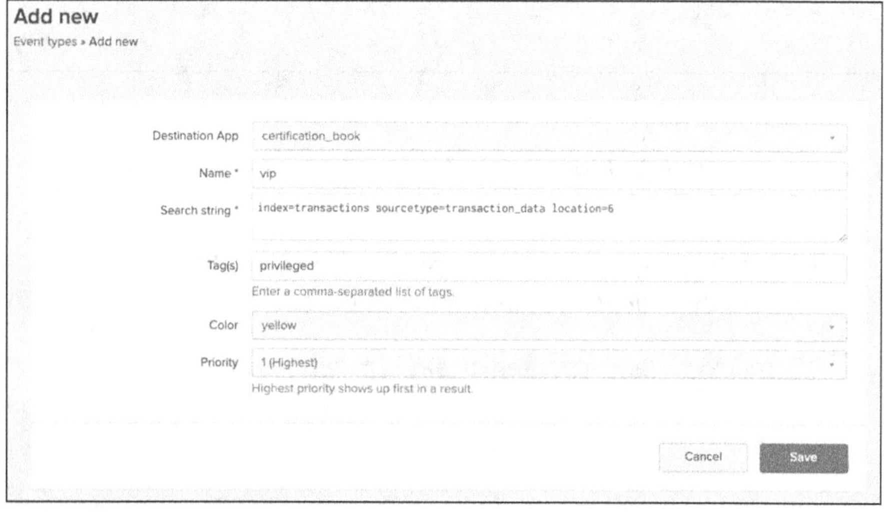

Figure 4-8. *Splunk tag using event type*

4. Click Save.

5. If you call all the events with a vip tag in the search
    bar, you find all events that are priority 1 and the
    color yellow:

```
index=* eventtype=vip earliest=0
```

All the events that are priority 1 and the color yellow, as stated in the
search query, are shown in Figure 4-9.

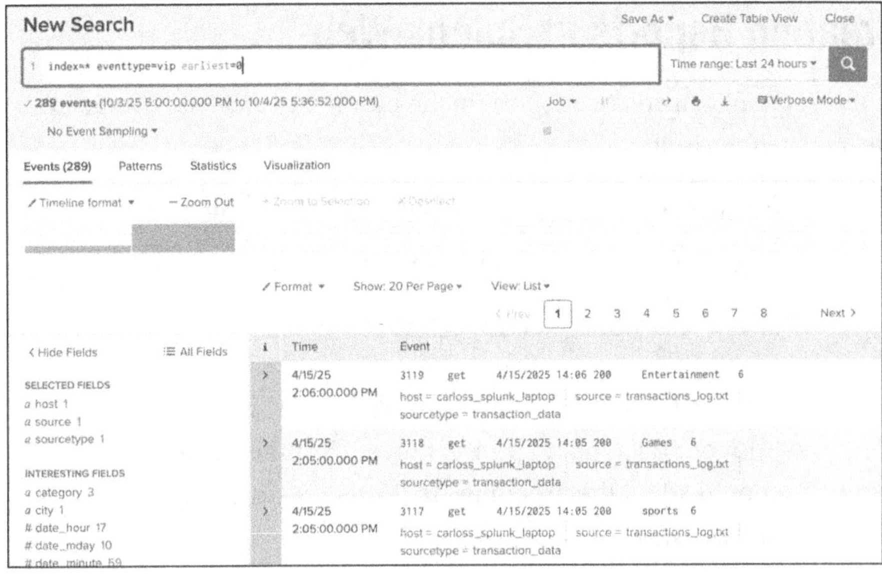

Figure 4-9. *Splunk tag using event type*

Set the priority and color based on the specific event's importance. If
you find a particular event that is more important, give it a higher priority
(the opposite for a low priority event).

The following section covers the process of reporting in Splunk.

Reporting in Splunk

Reports are saved searches that contain defined information represented through events, tables, charts, visualizations, and so forth to be reused. Reports in Splunk can be configured to perform versatile operations and can be converted into PDF format and sent as an email. Reports can also be scheduled in Splunk.

Let's start learning how to create reports in Splunk Web.

Creating Reports in Splunk Web

In this section, you create a report on the number of received events from the transactions index.

1. Navigate to the certification_book app in Splunk.

2. Type the following search command:

    ```
    index=transactions ip=* | stats count by ip
    ```

3. In Time Range Picker, select All Time. Note that this is for our data, but in production environments, you should adapt the timerange to the use case (last 24h, last week, etc.)

4. Go to Visualization and select Column Chart. The output is shown in Figure 4-10.

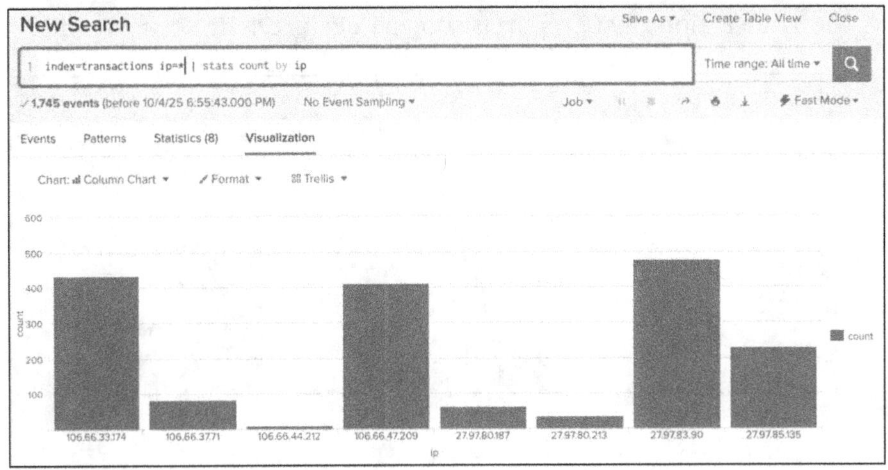

Figure 4-10. *Splunk visualization*

5. In the upper right corner, click Save As.

6. In Title, type **Transactions by IP**. In Content, select
    table + chart (see Figure 4-11).

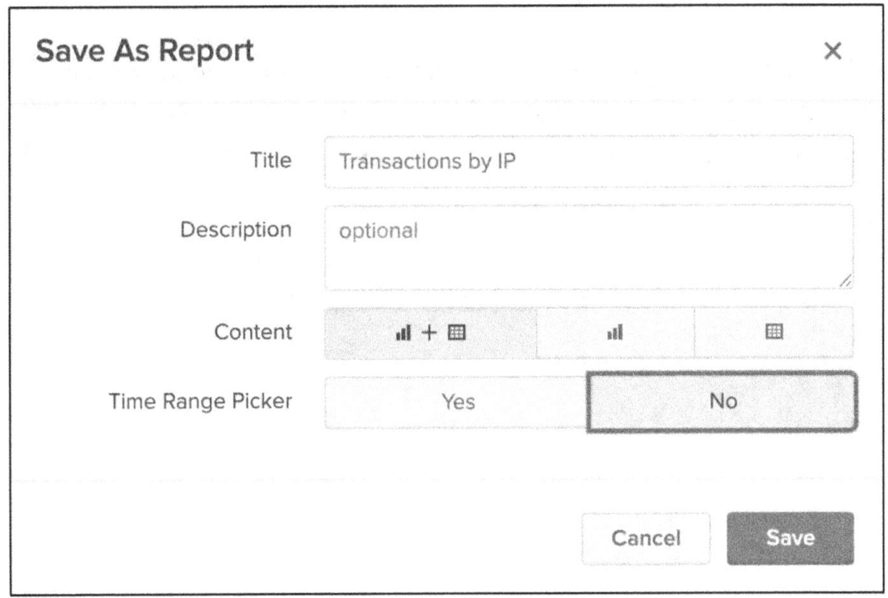

Figure 4-11. *Splunk report*

The Transactions by IP report is shown in Figure 4-12.

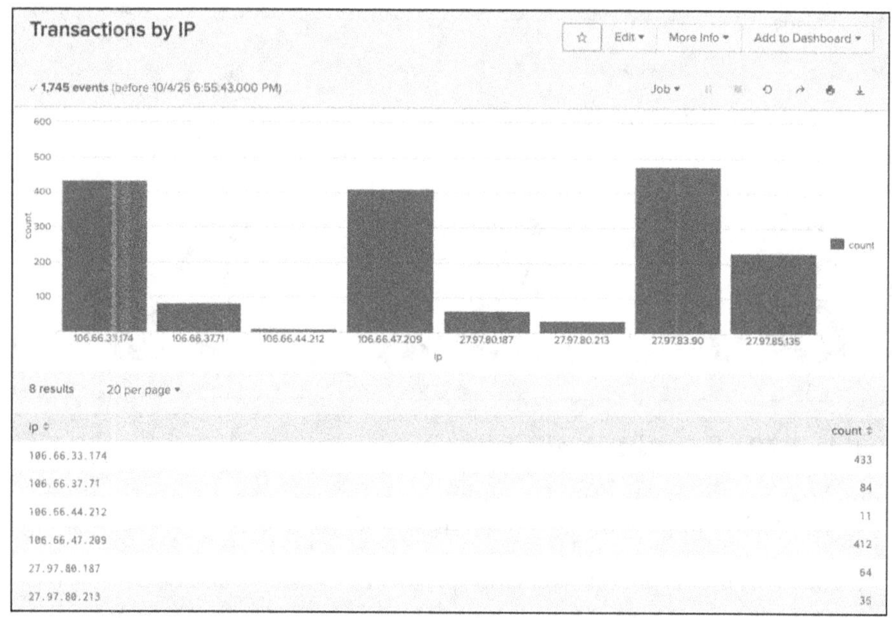

Figure 4-12. *Splunk Test1 report*

You can also create a report by going to Settings ➤ Searches ➤ Reports and Alerts ➤ Create a New Report. In this report, you may also get two extra fields: the earliest time and the latest time. Figure 4-13 represents an example.

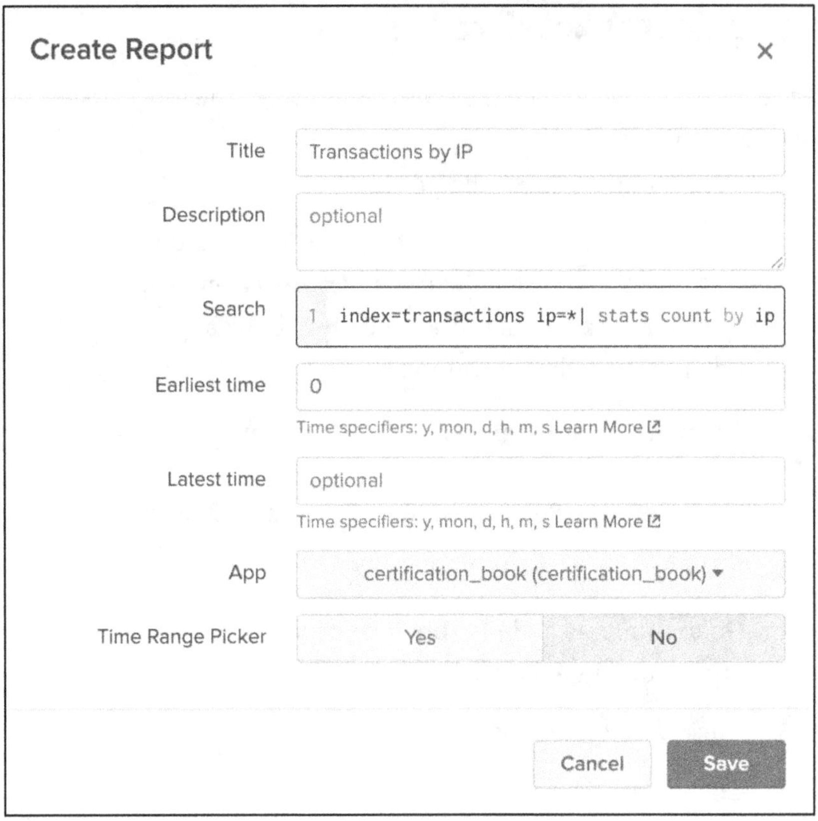

Create Report
×
Title
Transactions by IP
Description
optional
Search
1   index=transactions ip=*| stats count by ip
Earliest time
0
Time specifiers: y, mon, d, h, m, s Learn More
Latest time
optional
Time specifiers: y, mon, d, h, m, s Learn More
App
certification_book (certification_book) ▼
Time Range Picker
Yes
No
Cancel
Save

Figure 4-13. *Report in Splunk*

Next, you learn about report acceleration in Splunk.

Report Acceleration in Splunk

Report acceleration in Splunk means the platform systematically runs
a background process that builds a data summary based on the results
returned by the report. The user will consume the data summary rather
than the full index, which will increase the search return results. It usually
applies to large data sets which take a long time to run the report.

Creating Report Acceleration

In this section, you accelerate the report to discover the top ports used in index="transactions".

Follow these steps (also see Figure 4-14):

1. Create a report to learn the top ports for the index = transactions. Name the report **Transactions Top Ports**. In-app, select the certification_book app. Refer to the following query:

   ```
   index=transactions port=* |  top port
   ```

2. Go to Settings. Select Searches, Reports, and Alerts.

3. In App, select certification_book.

4. Go to the Transactions Top Ports report. Click Edit and go to Edit Acceleration.

5. Click Accelerate report.

6. Click Next.

7. In Summary Range, select All Time.

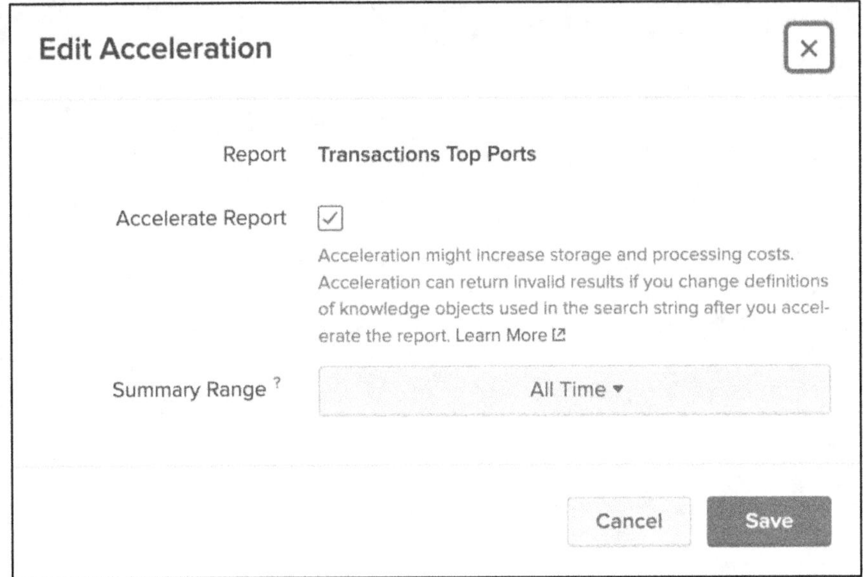

Figure 4-14. *Acceleration page*

Do the following to check the report summary:

1.   Go to Settings.

2.   Select Report Acceleration and summary.

3.   When you click Summary Id, you get content like
     event size on disk, summary range, buckets, chunks,
     and so forth (see Figure 4-15).

Summary Details
Report Acceleration Summaries » Summary Details

Summary: 482164195d3dfa24

Summary Status

Complete   Updated: 1m ago

Actions

Verify   Update   Rebuild   Delete

Reports Using This Summary

Search name	Owner	App
Transactions Top Ports	admin	certification_book

Details
Learn more.

Summarization Load	0.0007
Access Count	0 Last Access: Never
Size on Disk	1.32MB
Summary Range	All Time
Timespans	10min, 1d, 1h
Buckets	3
Chunks	169

Figure 4-15. Sample summary

Report acceleration in Splunk uses automatically created summaries to speed the process, but there is a cost associated with it because the number of accelerated reports accounts for concurrent searches and could lead to increase the usage in the indexer and search head.

Scheduling a Report in Splunk

A scheduled report is a predefined report that runs in fixed intervals. Once a report is triggered, you can define its actions into various steps, like sending a report summary by email, reporting the results through a lookup table, using webhooks, or logging and indexing searchable events. In this section, you schedule a report and run it every Monday at 10:00 with the highest priority.

To do this, refer to the following steps:

1. Go to Settings. Select Searches, Reports, and Alerts.

2. In App, select certification_book. Go to Transactions by IP report. Click Edit and select Edit Schedule.

3. In Edit Schedule, select Run every week on Monday at 10:00, and set priority to Highest (see Figure 4-16).

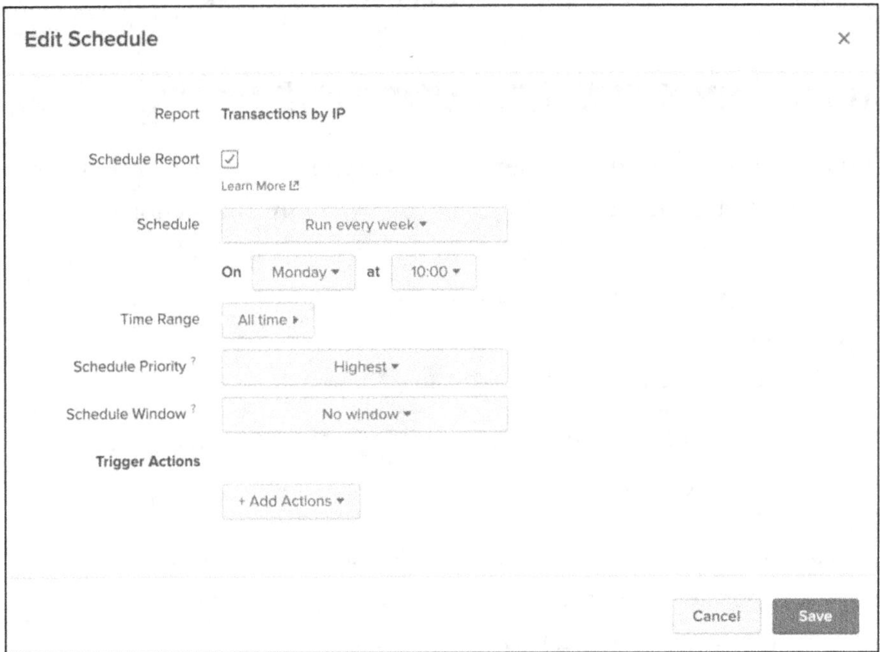

Figure 4-16. *Schedule page*

To edit a scheduled report, you need to get write access from the Splunk admin.

That completes the process of creating and reporting in Splunk Web. The following section discusses alerts.

Alerts in Splunk

Alerts are very important in organizations, since they use saved searches to proactively monitor machine data, triggering an action when user-defined conditions are met, such as when the number of failed login attempts exceeds a threshold. They trigger when the saved search meets a certain condition.

Let's start with learning how to create alerts in Splunk Web.

Create Alerts in Splunk Using Splunk Web

Alerts in Splunk monitor and alert when search results meet certain conditions. Once the alert is triggered, you can define the actions. In this section, you create an alert that runs on the first day of the month at 0:00. You want to know when the top port's event count exceeds or is equal to 400.

Follow these steps to perform this task:

1. Go to Settings. Select Searches, Reports, and Alerts.

2. Select New Alerts.

3. Set the title as Transaction Ports Exceeded Limits and the app name as certification_book. Select run every month on day 1 at 0:00. Refer to Figure 4-17.

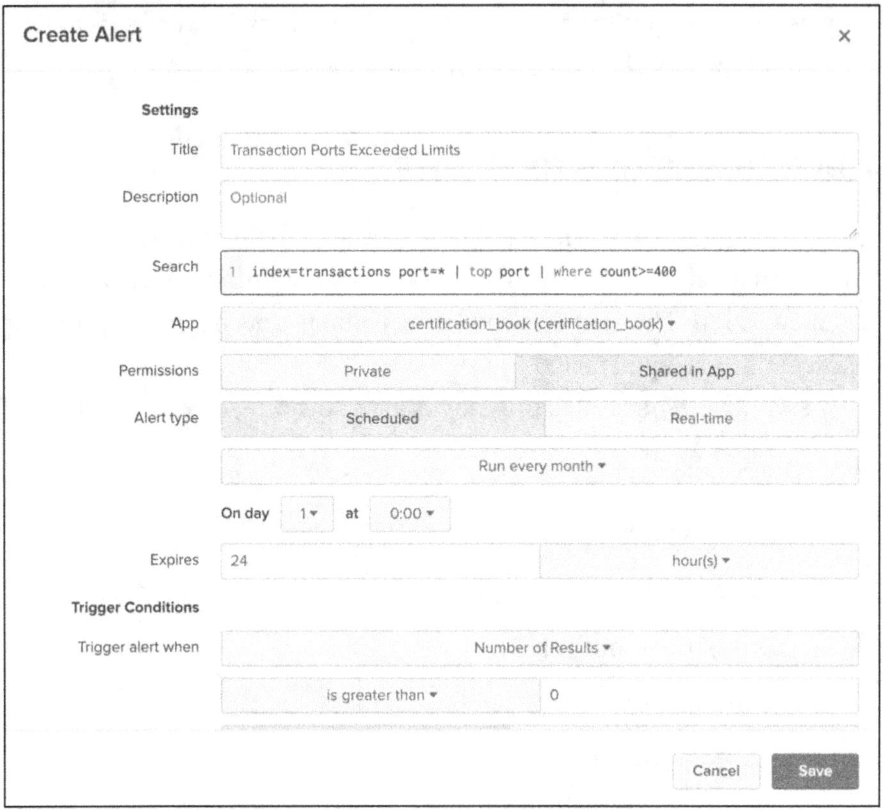

Figure 4-17. *Alert in Splunk*

4. Create a trigger in Splunk, as shown in Figure 4-18.
    Set the severity to Low.

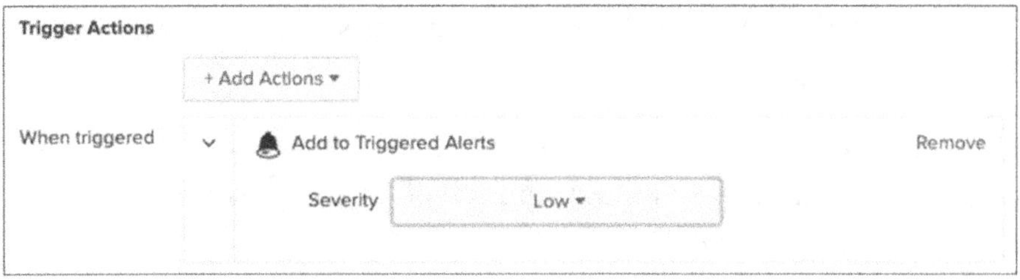

Figure 4-18. *Trigger in alert*

By default, alert visibility depends on permissions. Power users and admins have explicit rights to write alerts in Splunk.

Cron Expressions for Alerts

You can use cron expressions for alert scheduling in Splunk. In this section, you create a cron expression that has the alert run twice per day, at 12 AM and PM. The alert lets you know when the event count for the top port exceeds or is equal to 500.

The syntax for the cron expression is as follows:

```
<M H D m DW>
```

- Minutes: 0–59

- Hours: 0–23

- Day of the month: 1–31

- Month: 1–12

- Day of the week: 0–6 (where 0 = Sunday)

The following explains how to edit this alert:

1. In the certification_book app, go to Settings.

2. Select Searches, Reports, and Alerts.

3. Select Alerts.

4. Edit the alert name to Transaction Ports Exceeded 500.

5. In the **Run every month** drop-down menu, select **Run on Cron Schedule**, and set the cron expression as shown in Figure 4-19.

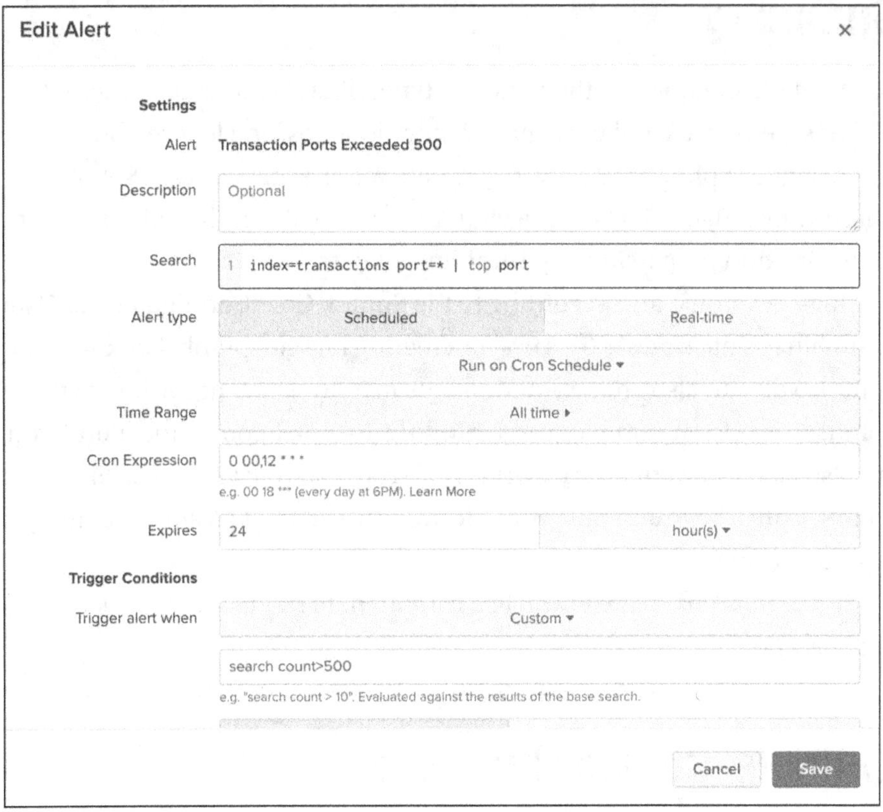

Edit Alert

Settings

Alert   Transaction Ports Exceeded 500

Description   Optional

Search   `1   index=transactions port=* | top port`

Alert type   Scheduled   Real-time

Run on Cron Schedule ▾

Time Range   All time ▸

Cron Expression   `0 00,12 * * *`

e.g. 00 18 *** (every day at 6PM). Learn More

Expires   24   hour(s) ▾

Trigger Conditions

Trigger alert when   Custom ▾

`search count>500`

e.g. "search count > 10". Evaluated against the results of the base search.

Cancel   Save

Figure 4-19. *Alert page*

6.   Click Save.

You have created an alert that runs twice per day (12 AM and PM) using the following cron expression:

```
0 00,12 * * *
```

Summary

You studied lookups and their various types, like CSV lookups, external lookups, geospatial lookups, and KV Store lookups. You learned how to create tags in Splunk and how to generate reports and alerts in Splunk. These knowledge objects play a vital role, especially for the Splunk admin managing and processing large amounts of data.

Tags in Splunk, and according to the Splunk Core Certified Power User exam blueprint, **Module 6—10%** for creating tags in Splunk. Lookups in Splunk, you can also create your lookups in Splunk and according to the Splunk Core User exam blueprint, **Module 7—6%**. Reports and alerts, you are also able to create your report and alerts in Splunk and according to Splunk Core User exam blueprint **Modules 6 and 8—11%** for scheduling reports and alerts.

In the next chapter, you will learn about data models, Pivot, event actions, and CIM.

Multiple-Choice Questions

    A.   Select all options that are lookup types.

        1. CSV lookup

        2. KV Store lookup

        3. External lookup

        4. None of the above

    B.   In Splunk lookups, field values are case-insensitive by default

        1. True

        2. False

C.   In Splunk, tags can use event types. You can set the priority and color based on an event's importance.

   1. True

   2. False

D.   If I am _____ , I can edit a scheduled report.

   1. A user

   2. A power user

   3. An admin

   4. All of the above

E.   To create an alert, you need to be____. (Select all options that applies.)

   1. A user

   2. A power user

   3. An admin

   4. edit_user

F.   To create an alert in Splunk using .conf edit _____.

   1. savedsearches.conf

   2. props.conf

   3. alert.conf

   4. transformation.conf

G. You can access a tag in Splunk using ____. (Select all that apply.)

1. ::

2. =

3. *

4. &

Answers

- a. 1, 2, 3

- b. 2

- c. 1

- d. 2, 3

- e. 2, 3

- f. 1

- g. 1, 2, 3

References

- ```
https://help.splunk.com/en/splunk-enterprise/
alert-and-respond/alerting-manual/10.0/create-
alerts/use-cron-expressions-for-alert-
scheduling
```

- *Splunk Operational Intelligence Cookbook* by Josh Diakun, Paul Johnson, Derek Mock (Packt Publishing, 2018).
```

- `https://help.splunk.com/en/splunk-enterprise/
  manage-knowledge-objects/knowledge-management-
  manual/10.0/tags/about-tags-and-aliases`

- `https://help.splunk.com/en/splunk-enterprise/
  create-dashboards-and-reports/reporting-
  manual/10.0/report-management/create-and-
  edit-reports`

Data Models, Pivot, and CIM

Splunk's power isn't just in searching raw events; it is in shaping those events into fast, reusable analytics. In this chapter, you will learn how Data Models define the structure of your data, how Pivot lets anyone build rich tables and visualizations from those models without writing SPL, and how the Common Information Model (CIM) standardizes field names across diverse sources.

We will start designing Data Models: choosing constraints, defining event datasets, adding fields (auto-extracted, eval, lookup-backed), and mentioning Data Model Acceleration to generate high-performance summaries. Then we'll use Pivot to turn those models into repeatable reports, before promoting them into dashboards. Additionally, we will check how workflow actions convert insights into next steps directly from an event—checking external sources or pivoting to new searches.

Finally, we will map our data to the CIM: applying tags and field aliases, so your data conforms to Splunk's shared schema (e.g., src). You'll see how CIM alignment unlocks out-of-the-box detections, searches, and visualizations—especially in Splunk Enterprise Security—while keeping your SPL portable and maintainable.

© Carlos Moreno Buitrago, Deep Mehta 2026

C. M. Buitrago and D. Mehta, *The Splunk Core User Study Companion*, Certification Study Companion Series, https://doi.org/10.1007/979-8-8688-2501-9_5

By the end of this chapter, you will be able to

- Design and accelerate a Data Model for a real dataset.

- Build Pivot reports and visuals without SPL, then operationalize them.

- Create and use workflow actions.

- Normalize diverse sourcetypes to the CIM and validate the mapping.

Exam tip   Know the differences between Data Models vs. CIM (structure vs. standard), when to enable acceleration, and how Pivot relates to tstats and accelerated datasets.

A data model is a hierarchical structure of datasets. An event action is a highly configured knowledge object that communicates among the fields in Splunk Web and other configurable internet sources. A CIM can normalize data using field names and event tags to get events from different data sources. By the time you complete this chapter, you will have embraced 30% of the Splunk Core Certified Power User exam blueprint.

Understanding Data Models and Pivot

Data models are **hierarchical datasets that enable Pivot users to create reports and dashboards** without writing complex commands in SPL. When generating a report, data models and their respective datasets are encoded by the organization's knowledge managers or Splunk's admin.

The events could fit in one or multiple data models, to use the same field context across the dataset.

Let's discuss the datasets and data models to understand the SPL better.

Datasets and Data Models

Data models are composed of parent datasets and child datasets that are arranged hierarchically. Each child dataset inherits constraints and fields from its parent dataset, besides having its own (a concept similar to inheritance in programming).

A dataset is an **assemblage of data** that is defined and maintained for a specified purpose. It is depicted in a table comprising fields in columns and their values in cells to be used in visually rich Pivot reports and to extend themselves during the search.

Datasets include auto-extracted fields, eval expressions, lookups, regular expressions, and geo IP fields.

Creating Data Models and Pivot in Splunk

This section shows how data models are created and how Pivot recognizes a pattern of product sales requests from regions in the Eastern and Western United States.

Creating New Datasets

Follow these steps to create a new dataset:

1. Go to Settings and select the data models.
   Click Create a New Data Model. In Title, type
   Transactions. In App, select certification_book (see
   Figure 5-1).

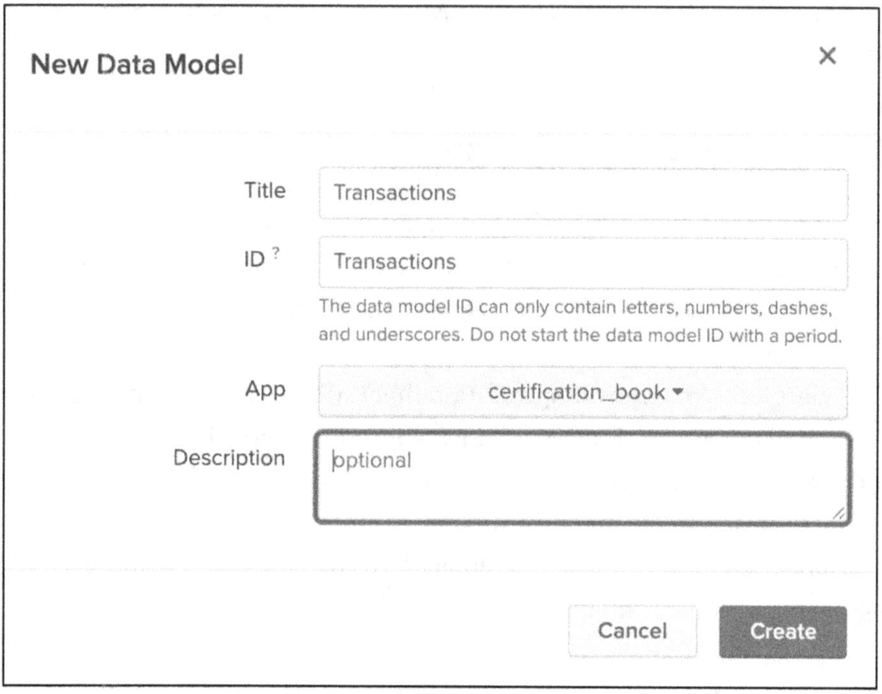

Figure 5-1. *Create Transactions Data Model*

2. Click Create. You are redirected to the dataset page.

3. Click Add Dataset, and select Root Event from the drop-down menu.

    The data model's Transactions page is shown in Figure 5-2.

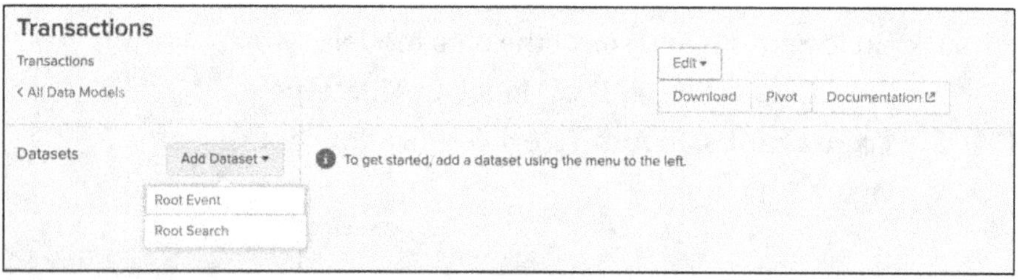

Figure 5-2. *Transactions Data Model*

4. In the Dataset Name, enter **All transactions**, and in the Constraints, type the following search command, and click in preview:

```
index=transactions
sourcetype=transaction_data
```

Figure 5-3 shows when the All transactions dataset is added.

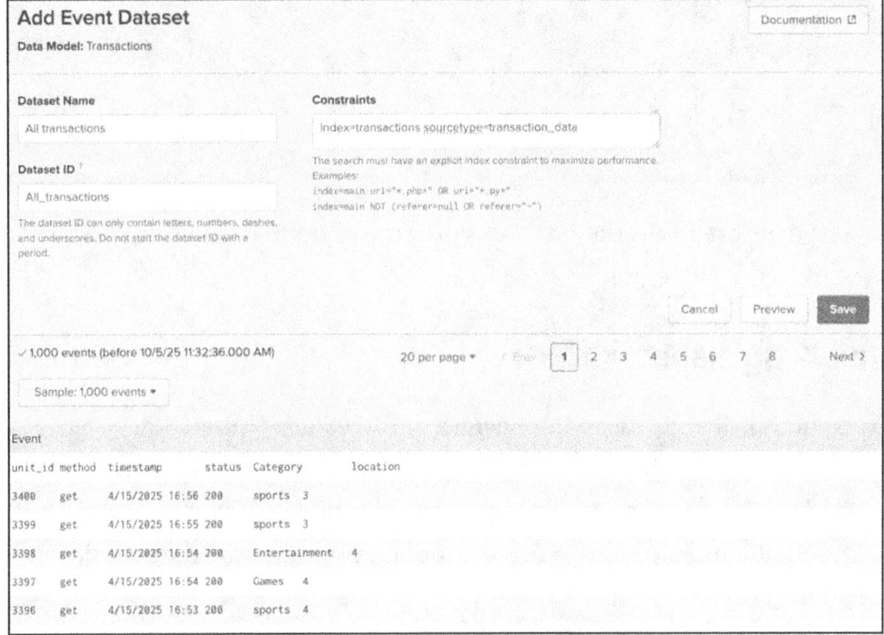

Figure 5-3. Data Model Transactions: All transactions dataset

5. Save the settings.

   Constraints in the dataset are essentially search terms.

6. Select Add Fields Auto-Extracted.

7. Select the following fields from the Auto-
    Extracted menu:

- category

- city

- location

- method

- status

- transaction_id

- unit_id

8. Save the settings.

You have created a dataset. Let's now move on to predicting sales patterns in the US regions.

Analyzing Sales Pattern

To predict sales patterns in the Eastern and Western United States, we have to use the city field we created in the previous chapter, using the match_locations automatic lookup. If you skipped that step, you can create the automatic lookup now or use the lookup option in the data model as shown in Figure 5-4.

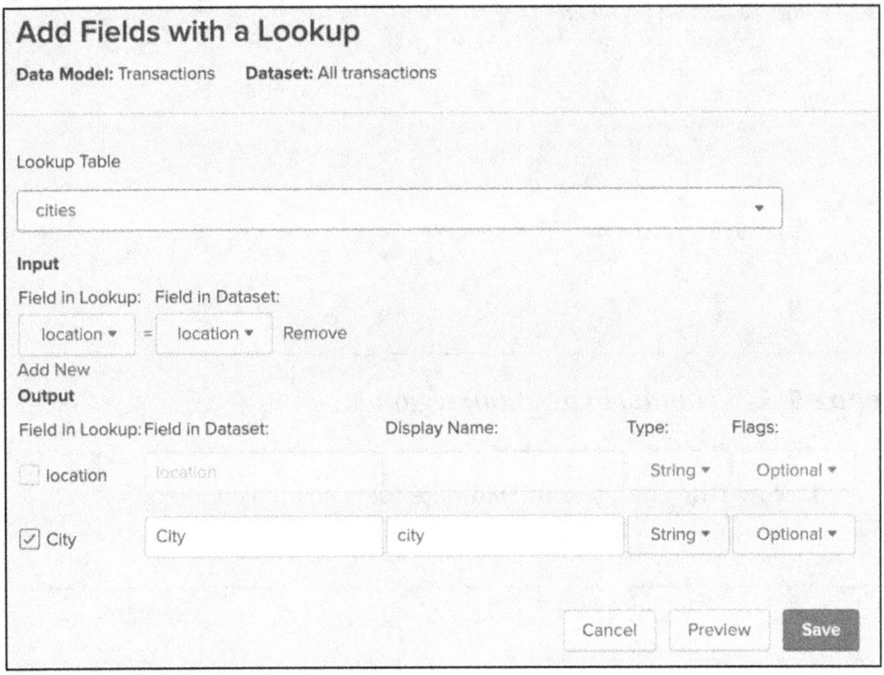

Add Fields with a Lookup

Data Model: Transactions   **Dataset:** All transactions

Lookup Table

| cities | ▼ |

Input

Field in Lookup: Field in Dataset:

location ▼ = location ▼   Remove

Add New

Output

Field in Lookup: Field in Dataset: Display Name: Type: Flags:

☐ location   location   String ▼   Optional ▼

☑ City   City   city   String ▼   Optional ▼

Cancel   Preview   **Save**

Figure 5-4. *Lookup field: City*

Now, you need to appropriately categorize the states into Eastern USA or Western USA. For this, you are using eval expressions:

1. Click Add Field and Evaluate Expression.

2. In Field Name, enter **region,** and in Display Name, enter **region**. In Eval Expression, type the following command:

```
case(city in("New york","Chicago","Boston",
"Washington D.C."),"Eastern USA", city in
("San Francisco","Seattle"),"Western USA")
```

These steps are shown in Figure 5-5.

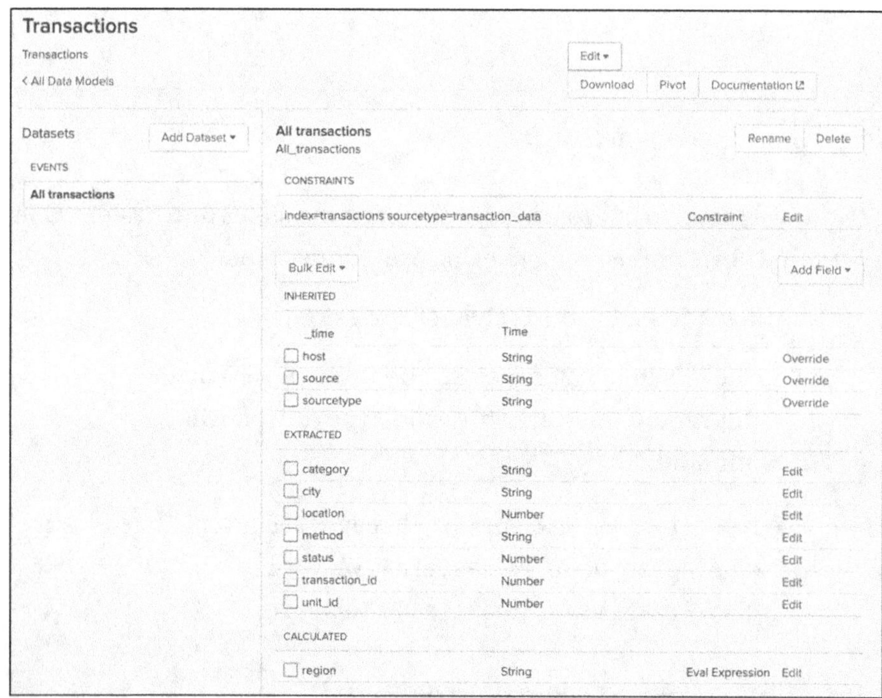

Figure 5-5. *Evaluate Expression: region*

3.   Save the changes, and validate that the dataset looks
      similar to Figure 5-6.

Figure 5-6. *All transactions dataset*

4. If you click Pivot, you can see the exact count for the product sales request obtained from the Eastern and Western USA.

5. Select Pivot as the All transactions dataset.

6. Split the rows as region, and split the columns as _time.

7. Go to the left side of the screen, select column chart, and in the range, select All Time. In the color section, update the periods to months.

8. Click on Save as, and save it as a dashboard. Choose a name for the dashboard.

In the next step, we will create two child datasets to split the data between regions:

1. Create a child dataset individually for the Western and the Eastern USA to obtain the product sales requests from each.

2. Click the All transactions dataset name. Go to Add Dataset, and select Child.

3. In Dataset Name, type **Western USA**. In the search command, type the following command.

```
region="Western USA"
```

Figure 5-7 shows the process of adding a child node named Western USA to the All transactions dataset.

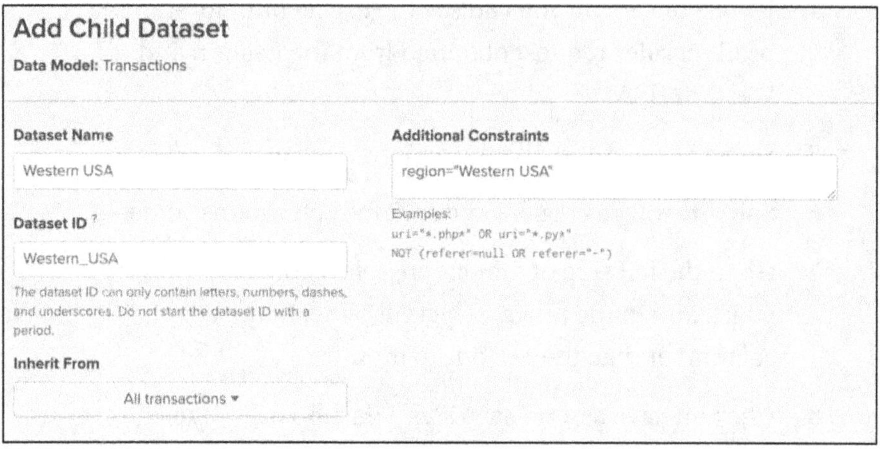

Add Child Dataset

Data Model: Transactions

Dataset Name

Western USA

Dataset ID ?

Western_USA

The dataset ID can only contain letters, numbers, dashes, and underscores. Do not start the dataset ID with a period.

Inherit From

All transactions ▾

Additional Constraints

region="Western USA"

Examples:
uri="*.php*" OR uri="*.py*"
NOT (referer=null OR referer="-")

Figure 5-7. *Western USA child dataset*

Child datasets inherit all constraints and fields from the parent dataset.

4.   Similarly, create another child for Eastern USA. Make the dataset name Eastern USA, and type the following search command. For more help, you can refer to the screenshot in Figure 5-8.

```
region="Eastern USA"
```

Figure 5-8 shows the process of adding a child node named Eastern USA to the All transactions dataset.

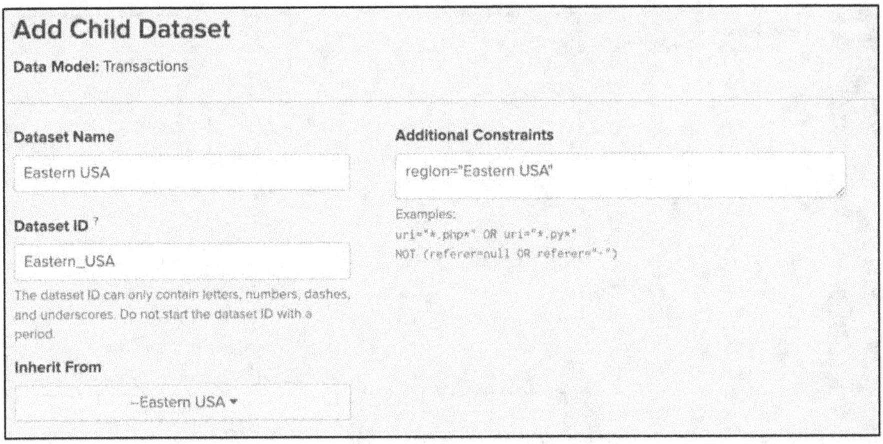

Add Child Dataset

Data Model: Transactions

Dataset Name

Eastern USA

Dataset ID [7]

Eastern_USA

The dataset ID can only contain letters, numbers, dashes,
and underscores. Do not start the dataset ID with a
period.

Inherit From

--Eastern USA ▾

Additional Constraints

region="Eastern USA"

Examples:
uri="*.php*" OR uri="*.py*"
NOT (referer=null OR referer="-")

Figure 5-8. *Eastern USA child dataset*

5. To get product sales requests from cities in the
   Western USA region, navigate to the Western USA
   child dataset.

6. Click Pivot for Dataset as Western USA.

7. In Split Rows, select city. In Split Columns,
   select _time.

   The screenshot in Figure 5-9 depicts the process
   of getting product sales requests from cities in the
   Western USA region over time.

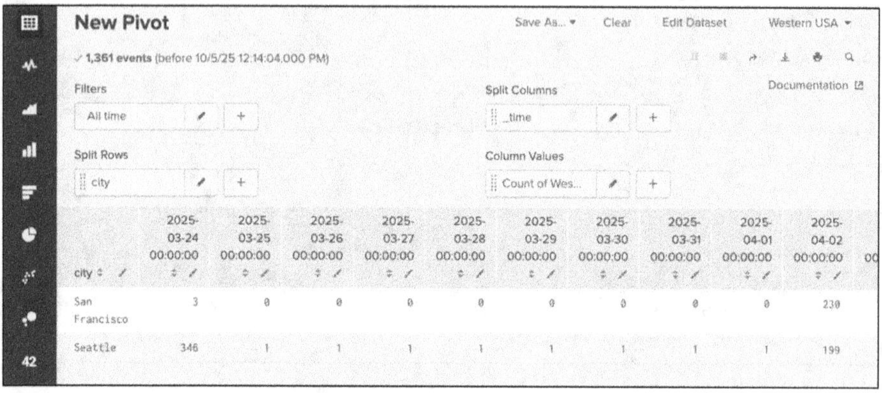

New Pivot

Save As... ▾   Clear   Edit Dataset   Western USA ▾

✓ **1,361 events** (before 10/5/25 12:14:04.000 PM)   Documentation ⧉

Filters: All time

Split Columns: _time

Split Rows: city

Column Values: Count of Wes...

city	2025-03-24 00:00:00	2025-03-25 00:00:00	2025-03-26 00:00:00	2025-03-27 00:00:00	2025-03-28 00:00:00	2025-03-29 00:00:00	2025-03-30 00:00:00	2025-03-31 00:00:00	2025-04-01 00:00:00	2025-04-02 00:00:00
San Francisco	3	0	0	0	0	0	0	0	0	230
Seattle	346	1	1	1	1	1	1	1	1	199

Figure 5-9. *Product sales request table: Western USA*

8.   If you do not need to split the data over time, you
     can remove the split option and get the total count.
     For our case, we want to get a total count per city
     in a pie chart format. Click the pie chart. In Range,
     select All time. You should see a pie chart similar to
     the one in Figure 5-10.

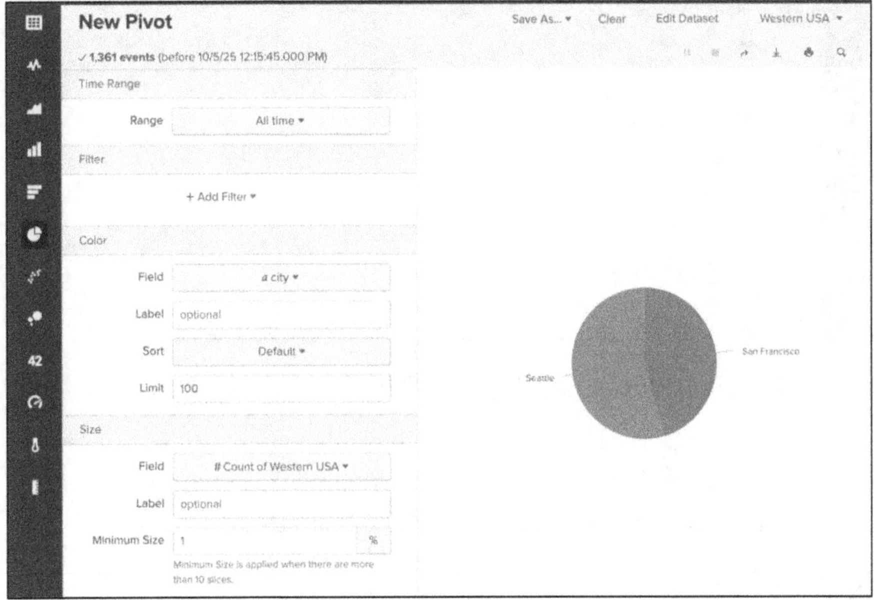

Figure 5-10. *Product sales request pie chart: Western USA*

9. Save the visualization in the dashboard you created in the previous steps.

10. To get product sales from states in the Eastern USA, navigate to the Eastern USA child dataset.

11. Click Pivot for Dataset as Eastern USA.

12. In Split Rows, select city.

The screenshot in Figure 5-11 depicts the process of getting product sales requests from cities in the Eastern USA region.

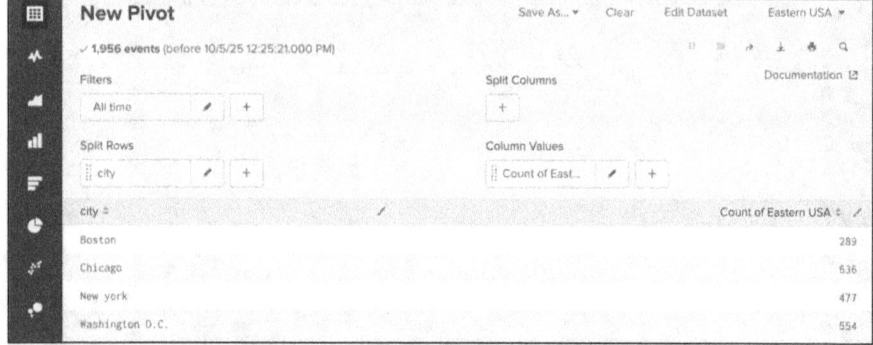

Figure 5-11. *Product sales Request Table: Eastern USA*

13.   Click pie chart. In Range, select All time. You should
      see a pie chart similar to the one in Figure 5-12.

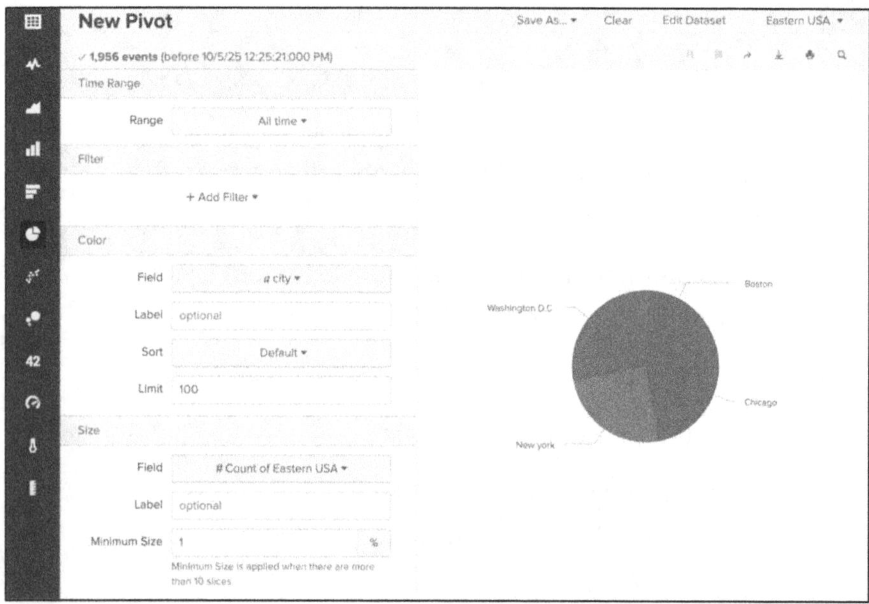

Figure 5-12. *Product Sales Request pie chart: Eastern USA*

14. Save the visualization in the dashboard you created
     in the previous steps, and check the results.

As the reports, you can optionally set data model acceleration to speed up the creation of Pivot tables and charts. To accelerate a data model, it must contain at least one root event dataset. Acceleration only affects these dataset types and datasets that are children of those root datasets.

You have created a dashboard with all the transactions by region in column format and two pie charts with the transactions split by city. Pivots allow you to check data and create dashboards in a simple way. You have completed the datasets and data models modules. After learning about dataset creation and practicing with a command-based application, you now move to event actions in Splunk.

Event Actions in Splunk

An event action, also known as a workflow, is considered a thoroughly configured knowledge object responsible for starting communication between a field in Splunk Web and other web resources. For instance, an event has IP as a field value, and company X has a webpage designed to check IP info. Here, the webpage gets the value of IP field, looks for IP info, and so forth. So, using an event action, you can generate POST requests and parse data into a specific URI.

In this section, you launch secondary searches for the IP field from the client_data.log source using event actions. You generate a request for the IP field in Splunk Web.

There are three types of workflows in Splunk:

- GET workflow actions

- POST workflow actions

- Search workflow actions

GET Workflow Actions

The GET workflow action is responsible for passing field values to HTML links, similar to an HREF tag. By clicking a HyperText Transfer Protocol (HTTP) link, the GET request allows you to pass a Splunk field value to a specific URI.

Defining a GET Workflow Action

The following are the steps to create a GET workflow in Splunk:

1. Go to Settings and select Fields.

2. Go to Workflow actions.

3. Click New to open a workflow action.

4. In the workflow action form, provide the following field name values and set an open link in New Window.

   ```
   App=certification_book
   Name=search_ip
   Label=search_ip
   apply only to the following fields=ip
   URI=https://www.virustotal.com/gui/ip-address/$ip$
   Link method=get
   ```

   Figure 5-13 shows the field values for a GET workflow action form.

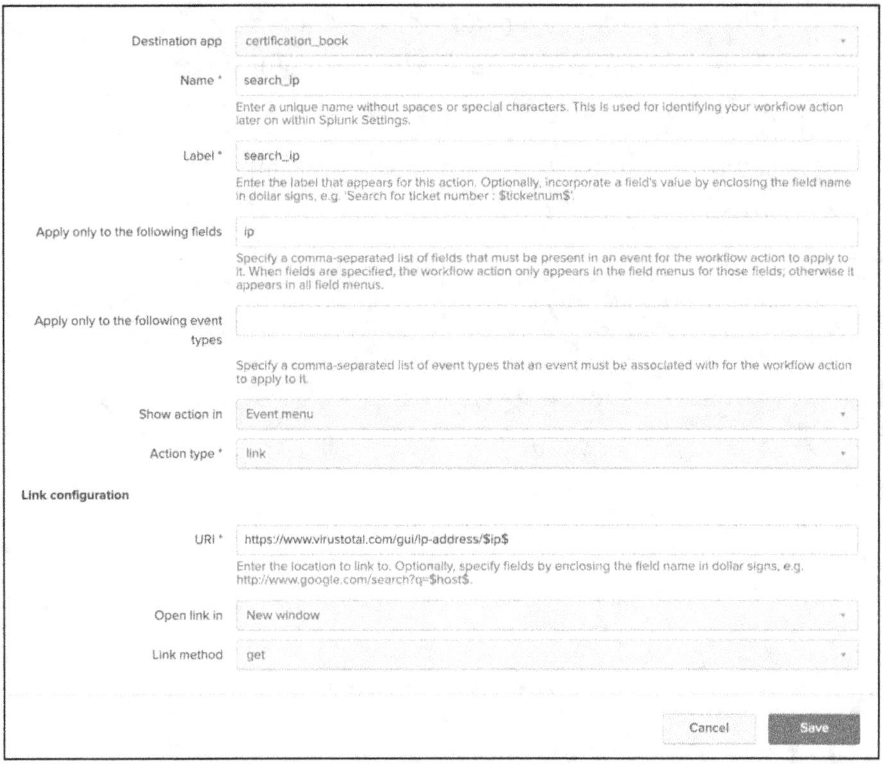

Figure 5-13. *search_ip workflow action*

The workflow action's name cannot have spaces or special characters.

5. Click Save.

6. To check the GET workflow action, go to the certification_book app, and type the following search command:

```
index=transactions sourcetype=client_data ip=*
earliest=0
```

7. Click any random event and click Event actions. There, you find search_ip as a link. Click the link; it performs an HTTP GET request to virustotal.com.

   Figure 5-14 shows the field values for a search workflow action form.

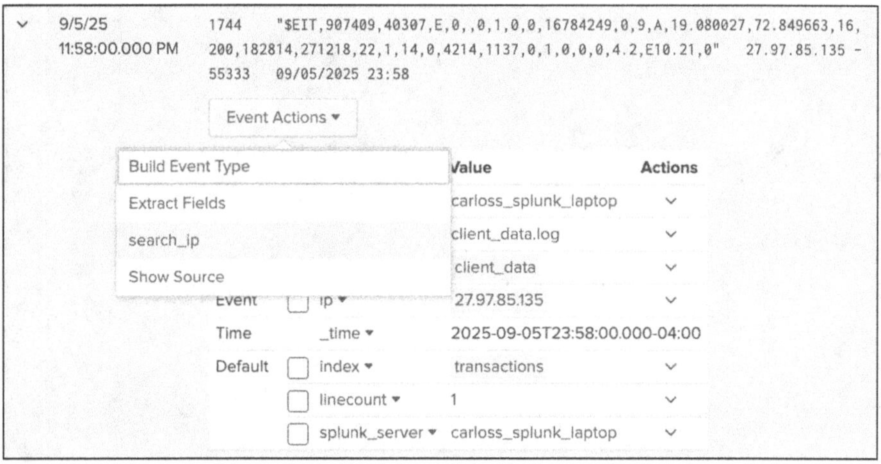

Figure 5-14. search_ip action in an event

Search Workflow Action

A search workflow action **generates dynamic searches in Splunk**. Instead of writing the entire SPL command for a particular field or determining the result for getting output, you can use a search workflow action for effective search results.

Defining Search Workflow Action

Here's an example of creating a search workflow action to get the total count of IP values in the _time parameter.

1. Go to Settings and select Fields.

2. Go to Workflow actions.

3. Click New to open a workflow action.

4. In the workflow action form, provide the following
    field name values and set an open link in
    New Window:

```
Name=timechart_by_ip
Label= Timechart by IP
apply only to the following fields=ip
show action in=Both
Action type=Search
Search string=index=transactions sourcetype=client_data
ip=$$ earliest=0 | timechart count by ip
Run in app=certification_book
```

Figure 5-15 shows the field values for the search
workflow action form.

Destination app	certification_book
Name *	timechart_by_ip
	Enter a unique name without spaces or special characters. This is used for identifying your workflow action later on within Splunk Settings.
Label *	Timechart by IP
	Enter the label that appears for this action. Optionally, incorporate a field's value by enclosing the field name in dollar signs, e.g. 'Search for ticket number : $ticketnum$'.
Apply only to the following fields	ip
	Specify a comma-separated list of fields that must be present in an event for the workflow action to apply to it. When fields are specified, the workflow action only appears in the field menus for those fields; otherwise it appears in all field menus.
Apply only to the following event types	
	Specify a comma-separated list of event types that an event must be associated with for the workflow action to apply to it.
Show action in	Both
Action type *	search
Search configuration	
Search string *	index=transactions sourcetype=client_data ip=ip earliest=0 | timechart count by ip
	Enter the search for this action. Optionally, specify fields as $fieldname$, e.g. sourcetype=rails controller=$controller$ error=*.
Run in app	certification_book
	Choose an app for the search to run in. Defaults to the current app.
Open in view	
	Enter the name of a view for the search to open in. Defaults to the current view.
Run search in	New window

Figure 5-15. timechart_by_ip workflow action

5. The search can run in a new window or the current window in the search workflow action. Keep in mind you can use post as the link method, depending on the requirements of the external resource.

6. Click Save.

7. To check the search workflow action, go to the certification_book app, and type the following search command:

```
index=transactions sourcetype=client_data ip=*
earliest=0
```

8. Click any random event and then on IP field actions.
    Here, you find the IP value as a link that generates a
    dynamic search in Splunk.

    Figure 5-16 shows the field values for the search
    workflow action form.

```
∨   9/5/25          1744     "SEIT,907409,40307,E,0,,0,1,0,0,16784249,0,9,A,19.080027,72.849663,16,200,182814,271
    11:58:00.000 PM   218,22,1,14,0,4214,1137,0,1,0,0,0,4.2,E10.21,0" 27.97.85.135 - 55333    09/05/2025 23:58
```

Type		Field	Value	Actions
Selected	✓	host ▾	carloss_splunk_laptop	∨
	✓	source ▾	client_data.log	∨
	✓	sourcetype ▾	client_data	∨
Event	☐	ip ▾	27.97.85.135	∨
Time		_time ▾	2025-0 Edit Tags	
Default	☐	index ▾	transa Timechart by IP	
	☐	linecount ▾	1	∨
	☐	splunk_server ▾	carloss_splunk_laptop	∨

Figure 5-16. *timechart_by_ip in an event*

Figure 5-17 shows the output of the search workflow
action that provides the total counts of a particular
IP on a time chart.

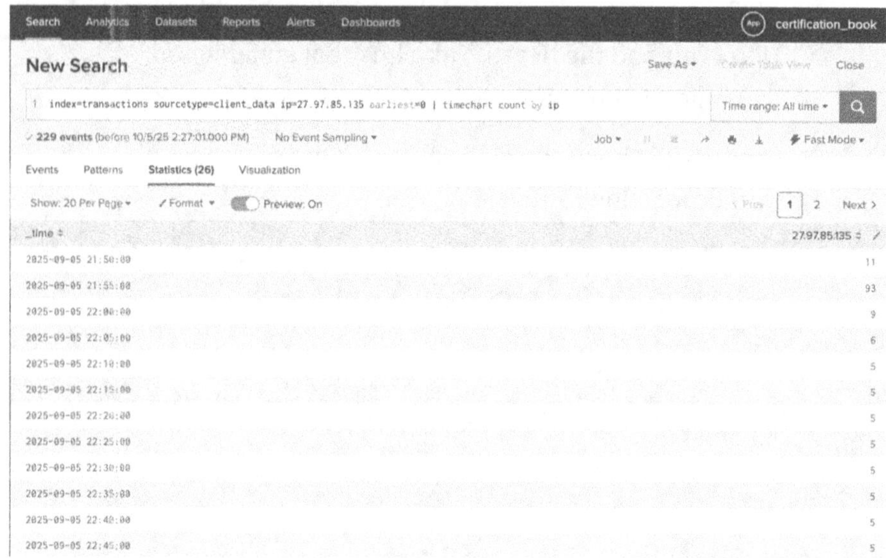

Figure 5-17. *Results of timechart_by_ip workflow action*

Common Information Model in Splunk

A Common Information Model (CIM) is a shared semantic model focused on extracting value from data. The CIM is implemented as an add-on that contains a collection of data models, documentation, and tools that support the consistent, normalized treatment of data for maximum efficiency at search time. In CIM, the data model comprises tags or a series of field names. They normalize data, using the same field names and event tags to extract from different data sources.

Defining CIM in Splunk

The following steps show how to configure CIM:

1. Go to Apps and Find More Apps.

2. Search by CIM. Install the "Splunk Common Information Model (CIM)" app.

3. Go to Settings and Data Models. You will see new data models assigned to the Splunk_SA_CIM app. They are a predefined collection of CIM data models. Figure 5-18 illustrates some of the new data models.

Data Models

Data models enable users to easily create reports in the Pivot tool. Learn More [2]

28 Data Models   App: certification_book (certification_book) ▼   Visible in the App ▼   Owner: Any ▼   filter   20 per page ▼

i	Title ▲	Type ⇕	⚡	Actions		App ⇕	Owner ⇕	Sharing ⇕
>	Alerts	data model	⚡	Edit ▼	Pivot	Splunk_SA_CIM	nobody	Global
>	Application State (Deprecated)	data model	⚡	Edit ▼	Pivot	Splunk_SA_CIM	nobody	Global
>	Authentication	data model	⚡	Edit ▼	Pivot	Splunk_SA_CIM	nobody	Global
>	Certificates	data model	⚡	Edit ▼	Pivot	Splunk_SA_CIM	nobody	Global
>	Change	data model	⚡	Edit ▼	Pivot	Splunk_SA_CIM	nobody	Global
>	Change Analysis (Deprecated)	data model	⚡	Edit ▼	Pivot	Splunk_SA_CIM	nobody	Global
>	CIM Validation (S.o.S.)	data model	⚡	Edit ▼	Pivot	Splunk_SA_CIM	nobody	Global
>	Data Access	data model	⚡	Edit ▼	Pivot	Splunk_SA_CIM	nobody	Global
>	Data Loss Prevention	data model	⚡	Edit ▼	Pivot	Splunk_SA_CIM	nobody	Global
>	Databases	data model	⚡	Edit ▼	Pivot	Splunk_SA_CIM	nobody	Global
>	Email	data model	⚡	Edit ▼	Pivot	Splunk_SA_CIM	nobody	Global
>	Endpoint	data model	⚡	Edit ▼	Pivot	Splunk_SA_CIM	nobody	Global

Figure 5-18. *CIM data models*

4. Open the Network Traffic data model. You will notice a lot of extracted and calculated fields. They will be used to extract the information from the traffic events, such as firewall logs. Most of the technologies have specific Splunk add-ons to automatically bring the necessary knowledge objects, auto-extract, and align the data source with CIM fields and data models. That is key if you want to use Splunk Enterprise Security, which relies on CIM data models.

5. In our example, we can align the client transaction
    IPs with CIM. Create a field alias for the IP field to
    get the src_ip, following the process explained in
    Chapter 3. You can use Figure 5-19 as an example.

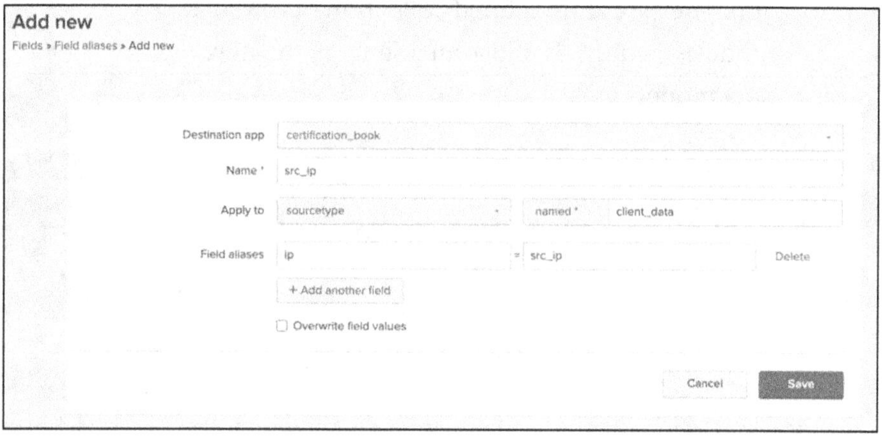

Figure 5-19. *src_ip field alias*

6. Create two new tags for the client_data events,
    called network and communicate. Check
    Figure 5-20 as an example.

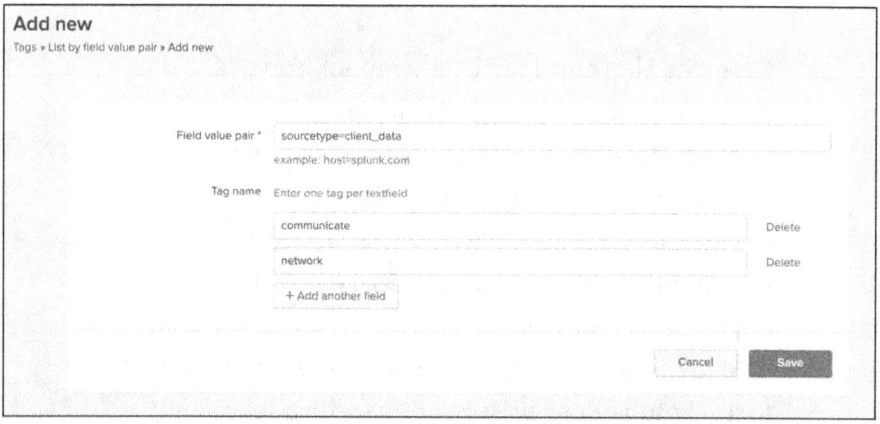

Figure 5-20. *CIM set up*

7. To set up CIM, go to Apps ➤ Manage Apps ➤ Splunk Common Information Model ➤ Set Up.

8. Find the Network Traffic Data Model section, and add transactions in the indexes allowlist as shown in Figure 5-21.

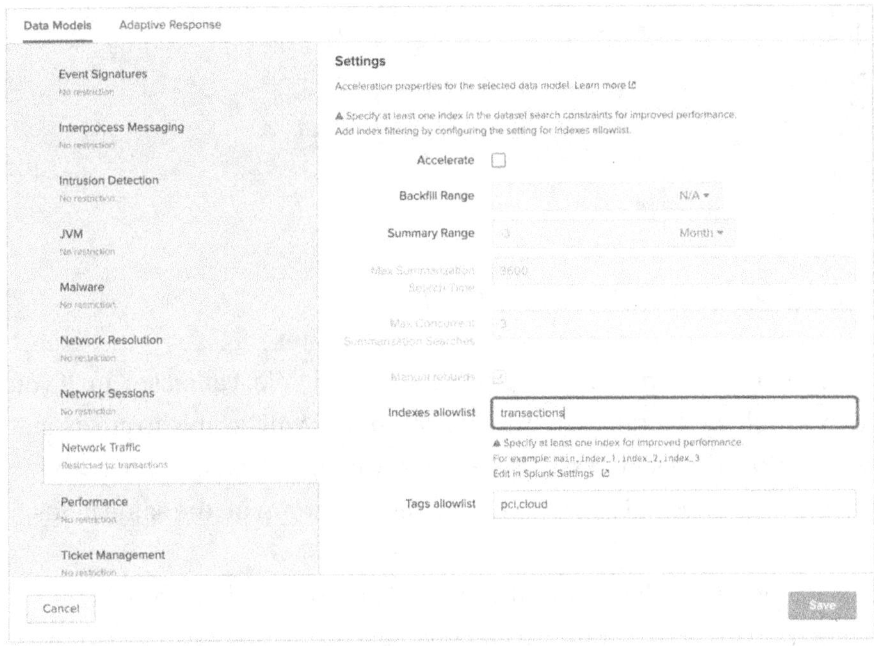

Figure 5-21. *CIM set up*

9. Go to Settings ➤ Data Models ➤ Network Traffic.

10. Run the Pivot model on the All Traffic dataset.

11. Split Rows by src_ip and check the results. Refer to Figure 5-22 to compare the results.

145

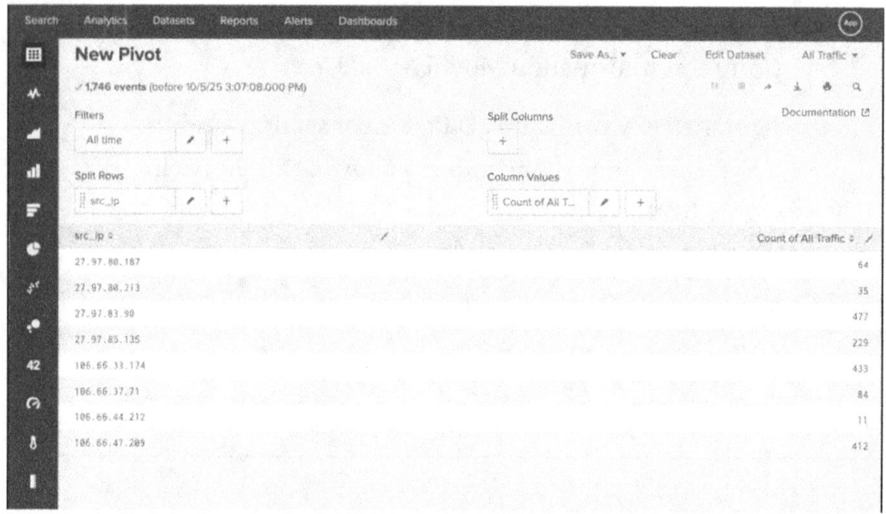

Figure 5-22. *Network Traffic Pivot for src_ip*

As you noticed, now we are using a standard field, called src_ip. If you add more allowed indexes in the CIM setup, you will be able to create stats from multiple data sources. CIM can be key in investigations where the analyst has to check different data sources and compare the same fields, such as src.

Data models included in CIM are configured with data model acceleration turned off, but you can enable it as needed on production environments.

You have reached the end of the chapter. I applaud you on successfully learning about datasets, data models, event actions, and CIM.

Summary

You learned that Splunk commands are simple and easy to understand. They are linked to HTTP and HTML without additional code or a high-level language and have common execution.

You have learned a hefty portion of workflow actions. Looking at the Splunk Core Certified Power User exam blueprint, you have become familiar with 10% of Module 9 (data models and Pivot), 10% of Module 8 (event actions), and 10% of Module 10 (CIM).

The next chapter covers knowledge managers and how to create dashboards in Splunk.

Multiple-Choice Questions

A. Select all options that are included as a type of events in a data model.

1. Events

2. Transactions

3. Searches

4. None of the above

B. What is the relationship between a data model and Pivot?

1. The data model provides datasets for Pivot.

2. Pivot and data models have no relationship.

3. Pivot and data model are the same thing.

4. Pivot provides a dataset for the data model.

C. Which fields are included in datasets?

1. Auto-extracted

2. Eval expression

3. Lookups

4. Regular expression

5. Geo IP fields

6. All the above

7. None of the above

D. Constraints in a dataset are essentially search terms.

1. True

2. False

E. Child datasets inherit all constraints and fields from the parent dataset.

1. True

2. False

F. Data models included in CIM are configured with data model acceleration turned on.

1. True

2. False

G. Select all options that include a type of workflow action in Splunk.

1. POST

2. GET

3. Search

4. None of the above

Answers

- a. 1, 2, 3
- b. 1
- c. 6
- d. 1
- e. 1
- f. 2
- g. 1, 2, 3

References

- https://docs.splunk.com/Splexicon:Dataset
- https://help.splunk.com/en/splunk-enterprise/manage-knowledge-objects/knowledge-management-manual/9.4/build-a-data-model/about-data-models
- https://help.splunk.com/en/splunk-cloud-platform/common-information-model/6.1/introduction/overview-of-the-splunk-common-information-model
- https://help.splunk.com/en/splunk-enterprise/manage-knowledge-objects/knowledge-management-manual/10.0/workflow-actions/about-workflow-actions-in-splunk-web

CHAPTER 6

Knowledge Managers and Dashboards in Splunk

SPL is only half the story. At scale, teams require reusable knowledge objects and purposeful dashboards to enable everyone to quickly verify data. This chapter shows how the Knowledge Manager function bridges human intent and machine execution by curating the objects that make searches simpler, faster, and consistent across your organization.

We'll explore how to govern the knowledge objects with ownership and permissions/scope (private, app, global).

Then we'll build compelling views with Splunk's two dashboard paradigms:

- **Dashboard Studio (JSON, visualization-first):**
  Modern layouts, rich visuals, and presentation control.

- **Classic dashboards (simple XML):** Highly interactive, token-driven forms with dynamic searches.
  Deprecated.

© Carlos Moreno Buitrago, Deep Mehta 2026

C. M. Buitrago and D. Mehta, *The Splunk Core User Study Companion*, Certification Study Companion Series, https://doi.org/10.1007/979-8-8688-2501-9_6

By the end of this chapter, you will be able to

- Curate and govern knowledge objects with the right scope and ownership.

- Create a dashboard using Dashboard Studio.

Understanding the Knowledge Manager's Role in Splunk

A knowledge manager is a person who provides centralized oversight and maintenance of knowledge objects in Splunk Enterprise. A knowledge manager creates, maintains, and modifies knowledge objects. Knowledge managers can reassign existing knowledge objects to another user, helping the team tackle day-to-day issues.

Five types of knowledge objects were covered in the previous chapters:

- Data interpretation (fields and field extraction)

- Data classification (event types and transactions)

- Data enrichment (lookups and workflow actions)

- Data normalization (tags and aliases)

- Data models

What exactly does a knowledge manager do? How can a knowledge manager help solve day-to-day issues? The following describes knowledge managers' responsibilities:

- They declare knowledge objects as global so that the entire organization can access them.

- They make knowledge objects available to apps in Splunk.

- They provide certain users access to a particular knowledge object.

- They manage orphaned knowledge objects.

- They restrict the read/write permission at the app level.

A knowledge manager builds data models for Pivot and ensures that the right person in the organization handles the knowledge objects. They also take care of normalizing a data event.

Now, let's discuss knowledge objects in greater detail.

Globally Transferring Knowledge Objects

Knowledge objects can be declared global to allow every user access to them in all the apps installed on the platform. For example, the sales, management, and production teams all have access to sales in the United States. This enables the sales team to recognize the sales graph, the management team to understand the organization's sales patterns, and the production team to know which products are profitable. By default, only a Splunk Power user or admin can modify knowledge objects.

To make knowledge objects global, follow these steps:

1. Go to Settings in Splunk Web, and select All Configurations.

2. Search for the knowledge object that you want to make global, and then click Permissions (see Figure 6-1).

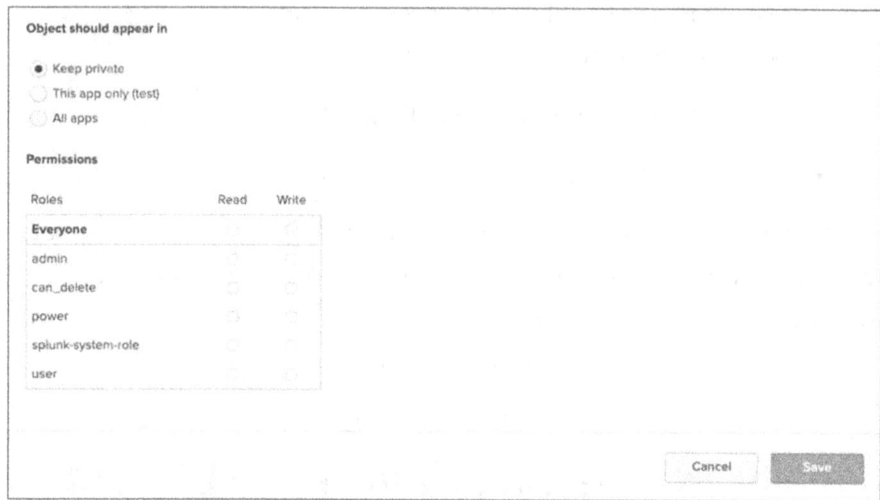

Figure 6-1. *Knowledge objects visibility and access management*

3. To make a knowledge object global, the object should be present in the All apps option. Select All apps for scope, then grant Read to the appropriate roles (see Figure 6-2).

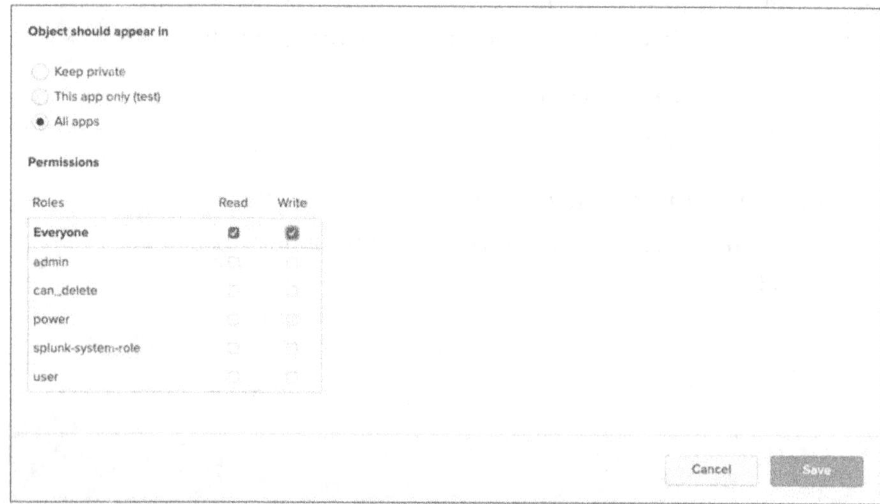

Figure 6-2. *Knowledge objects global visibility*

Enabling Knowledge Object Visibility

The knowledge objects related to sensitive information or hierarchical issues have restricted visibility. Not everyone in your organization has access to all the data. For example, the sales, management, and production teams all have access to data relating to Eastern and Western USA sales, but the organization's Supply Chain Management (SCM) team does not need this report. You can restrict object visibility to a specified audience.

To make a knowledge object selectively accessible, follow these steps:

1. Go to Settings and select All Configurations.

2. Search for the knowledge object that you want, and then click Permissions. I want src_ip to be visible only to users who have access to the certification_book app.

3. To make knowledge objects visible at app scope objects should appear in a particular app. Set the object to this app only, grant Read to the intended app users (see Figure 6-3).

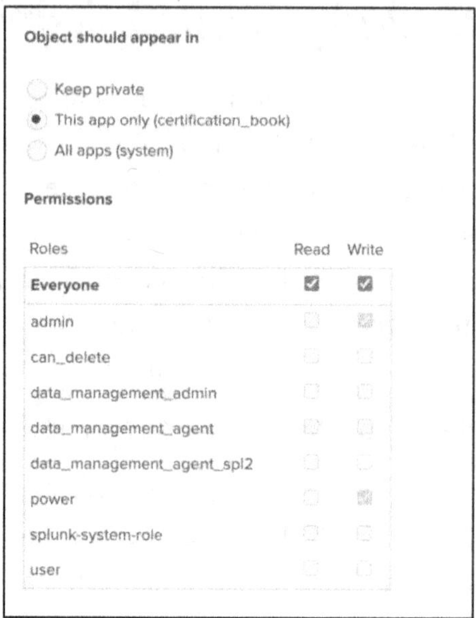

Figure 6-3. *Knowledge objects visibility for src_ip field in the certification_book app*

Restricting Read/Write Permissions on an App

Read/write permissions on an app are a knowledge object restriction as well. If you want users to only read objects and make no edits to them, follow these steps:

1. Go to Settings and select All configurations.

2. Search for the knowledge object, and then click Permissions. For example, src_ip will be a read-only object, visible in all the apps, and only the admin will be able to modify the KO.

3. In Permissions, change Read to everyone, and Write
    roles to admin (see Figure 6-4).

Object should appear in

Keep private

This app only (certification_book)

All apps (system)

Permissions

Roles	Read	Write
Everyone	☑	☐
admin	☐	☑
can_delete	☐	☐
data_management_admin	☐	☐
data_management_agent	☐	☐
data_management_agent_spl2	☐	☐
power	☐	☐
splunk-system-role	☐	☐
user	☐	☐

Figure 6-4. Knowledge objects access management

The Keep private option allows you to test the KO before it is added to
production or by the knowledge manager.

Orphaned Knowledge Objects

When employees leave an organization, the knowledge objects that were
linked to them are deactivated; however, the links to these objects remain.
These objects are called *orphaned knowledge objects*. Objects without
a legitimate owner are a hindrance in a Splunk environment, and the
scheduled report cannot trace and report them because the owner is no

longer available. To get around this, Splunk provides various methods to detect orphaned knowledge objects:

- It runs Monitoring Console health checks.

- It uses the Reassign Knowledge Objects page in Settings.

- It reassigns a single knowledge object to another owner.

Run a Monitoring Console Health Check

If you have admin role access in Splunk Enterprise, you can use the Monitoring Console. It is an app to check the overall status of your Splunk platform, and one of the features is a health check that detects orphaned scheduled searches, reports, and alerts. It tells you the number of knowledge objects that exist in your system. For this example, a user has been created, a test report has been assigned, and the user has been deleted. The following steps run the health checks:

1. Go to Settings and Monitoring Console. Click Health Check.

2. Filter by category=Splunk Miscellaneous

3. Click Start.

4. You shouldn't see an error in **Orphaned scheduled searches**, but it is the result of the scenario we created for this example (see Figure 6-5).

4	1	0	0	1	2
ALL	ERROR	WARNING	INFO	SUCCESS	N/A

Check ⇕	Category ⇕	Tags ⇕	Results ⇕
Orphaned scheduled searches	Splunk Miscellaneous	configuration, search, searches_skipped	⚠ One or more scheduled searches are orphaned, meaning that they are no longer associated with valid owners. The scheduler will not run orphaned scheduled searches.
Integrity check of installed files	Splunk Miscellaneous	configuration, installation	☺ This health check item was successful.
Excessive physical memory usage	Splunk Miscellaneous	resource_usage	☺ This health check item is not applicable.
KV Store status	Splunk Miscellaneous	kv_store	☺ This health check item is not applicable.

Figure 6-5. *Orphaned knowledge objects through monitoring console*

Using the Reassign Knowledge Objects Page in Settings

The Reassign Knowledge Objects page in Settings is one of the easiest methods for discovering if an orphaned knowledge object exists in a Splunk environment. Refer to the following steps:

1. Select Settings and go to All configurations.

2. Click Reassign Knowledge Objects.

3. Click Orphaned to filter it from the list.

The orphaned button also displays shared orphaned objects.

Reassigning a Knowledge Object to Another Owner

The Reassign Knowledge Objects page is used for reassigning a knowledge object to a new owner. You can reassign owned and orphaned knowledge objects. Follow these steps to do this:

1. Select Settings and go to All configurations.

2. Click Reassign Knowledge Objects.

3.   Find the object or objects that you want to reassign.

Figure 6-6 shows the Reassign Entry dialog box.

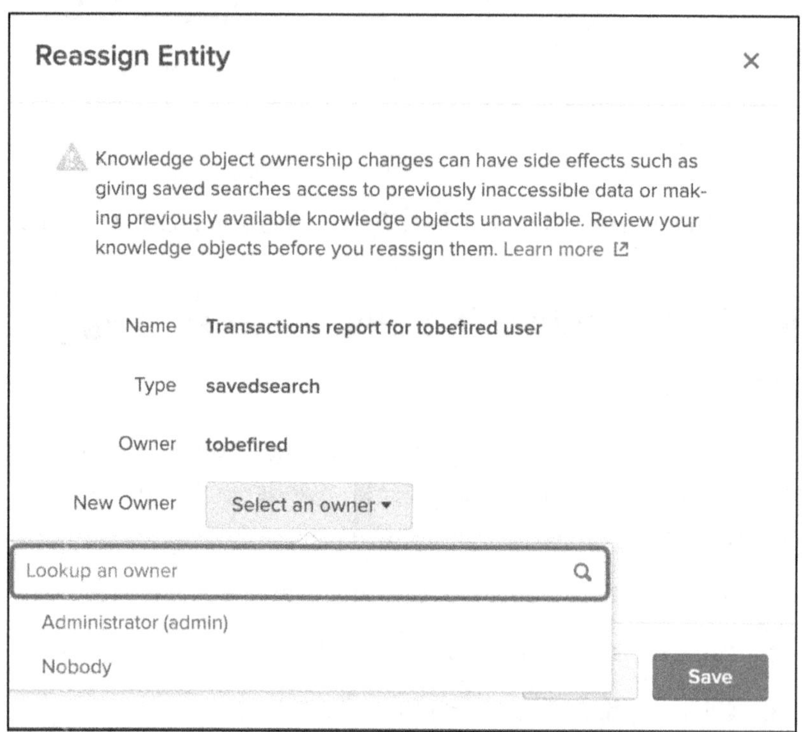

Figure 6-6. *Reassign knowledge objects*

Knowledge objects in Splunk are stored in macros.conf, tags.conf, eventtypes.conf, props.conf, transforms.conf, and savedsearches.conf.

At this stage of the chapter, you have already learned about the knowledge manager's role and how to regulate a knowledge object's visibility and restrict its read and write permissions.

Dashboards

Dashboards are designed to easily visualize data in Splunk through quick analyses and information summaries. The dashboard is convenient and straightforward because of the panels that constitute it. Panels include charts, tables, and lists directly linked to the reports.

There are two types of dashboards in Splunk:

- Classic dashboards

- Dashboard Studio

Classic vs. Studio

While both serve the purpose of visualizing data, they differ significantly in their underlying technology, layout flexibility, and feature sets.

Classic dashboard use simple XML as its source code. It is structured as a row and column layout, limiting design flexibility, but it supports third-party visualizations through the Splunk Custom Visualizations developer API, such as JavaScript.

Dashboard Studio utilizes JSON-formatted stanzas as its source code. It offers more flexible layout options, including absolute layout (free-form, pixel-perfect placement) and Grid layout (similar to classic but with enhanced features). Studio allows for greater visual customization with features like shapes, lines, icons, text boxes, images, background customization, and dynamic coloring options for visualizations like the single-value radial gauge.

At the time of writing this book, the classic dashboard is the most widely used one in organizations since it was the first dashboard option offered by Splunk. Many organizations are adopting Studio as Splunk implements new features. We will cover an example of each type of dashboard.

Classic Dashboards

In this section, you learn to create a classic dashboard for the transaction_ data sourcetype, where the user determines the total number of transactions on the web page, the total Eastern USA sales and total Western USA sales, the status of the HTTP transaction, the HTTP method, and the various product categories.

To create a dashboard for transaction_data sourcetype, you need the following eight queries:

- Total transaction requests on the web page

- Total transaction requests from Western USA

- Total transaction requests from Eastern USA

- Total successful transaction requests on the web page

- Total sales in Western USA cities

- Total sales in Eastern USA cities

- HTTP status code for the web page

- HTTP method for the web page

The solution for creating eight queries is explained next. If you're already confident with creating a dashboard, try making the reports without using the instructions. You can use the data model created in the previous chapter or use regular searches. To simplify the process, the searches will be added to each report idea.

Total Transaction Request on the Web Page

To create the dashboard, follow these steps:

1. Go to the certification_book app.

2. Dashboards ➤ Create New Dashboard.

3. Add the following information:

- **Dashboard Title**: Web Summary.

- **Description**: The following dashboard shows a general overview of the web.

- **Permissions**: Shared in App.

- **Dashboard type**: Classic Dashboards.

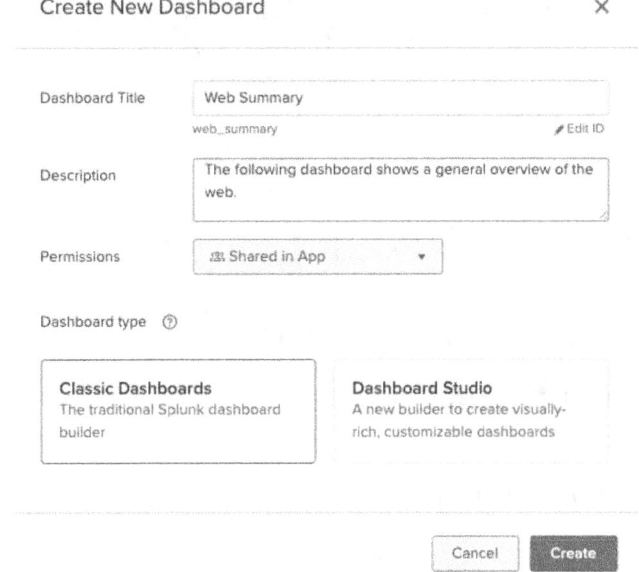

Figure 6-7. *Web summary: classic dashboard*

4. Click + Add Panel ➤ New ➤ Single Value

5. Complete the following information:

- **Time Range**: Use time picker ➤ All time

- **Content Title**: Total Transaction Request on the Web Page

- **Search String**:

  ```
  index=transactions sourcetype=transaction_data
  | stats count
  ```

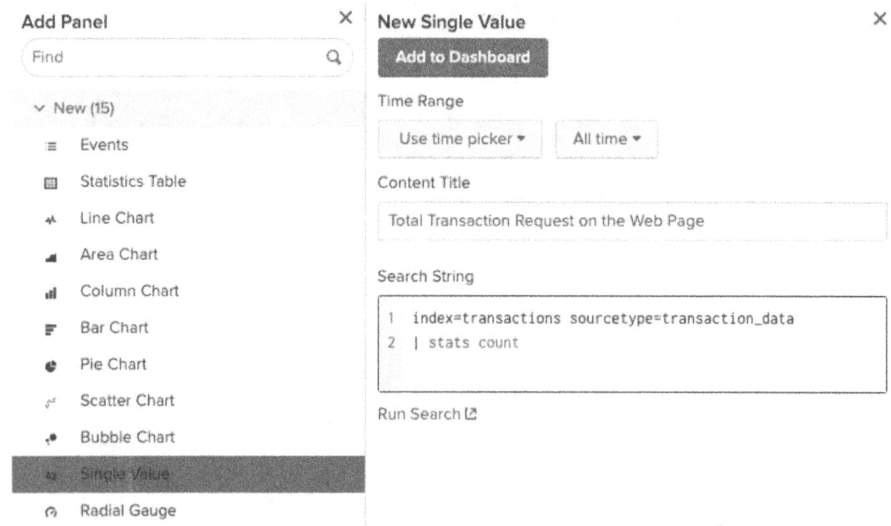

Figure 6-8. *Web summary: first panel*

Total Transaction Request from Western USA

To create this report, observe the following recommendations:

- **Panel Type**: Single value

- **Time Range**: All time

- **Search**:

```
index=transactions sourcetype=transaction_data
| eval region=case(city in("New york", "Chicago",
"Boston", "Washington D.C."), "Eastern USA", city
in("San Francisco", "Seattle"), "Western USA")
| search region="Western USA"
| stats count
```

Note   You noticed that the region could be a calculated field created for sourcetype=transaction_data. Feel free to create it and change this and the following queries by

```
index=transactions sourcetype=transaction_data
region=<region>
```

```
| stats count
```

Total Transaction Request from Eastern USA

To create this report, observe the following recommendations:

- **Panel Type**: Single value

- **Time Range**: All time

- **Search**:

  ```
  index=transactions sourcetype=transaction_data
  | eval region=case(city in("New york", "Chicago",
  "Boston", "Washington D.C."), "Eastern USA", city
  in("San Francisco", "Seattle"), "Western USA")
  | search region="Eastern USA"
  | stats count
  ```

Successful Transaction Requests on the Web Page

To create this report, observe the following recommendations:

- **Panel Type**: Single value

- **Time Range**: All time

- **Search**:

```
index=transactions sourcetype=transaction_data
status=200
| stats count
```

Successful Transaction Requests by Western US Cities

To create this report, observe the following recommendations:

- **Panel Type**: Pie Chart

- **Time Range**: All time

- **Search**:

```
index=transactions sourcetype=transaction_data
status=200
| eval region=case(city in("New york", "Chicago",
"Boston", "Washington D.C."), "Eastern USA", city
in("San Francisco", "Seattle"), "Western USA")
| search region="Western USA"
| stats count by city
```

Successful Transaction Requests by Eastern US Cities

To create this report, observe the following recommendations:

- **Panel Type**: Pie Chart

- **Time Range**: All time

- **Search**:

```
index=transactions sourcetype=transaction_data
status=200
| eval region=case(city in("New york", "Chicago",
"Boston", "Washington D.C."), "Eastern USA", city
in("San Francisco", "Seattle"), "Western USA")
| search region="Eastern USA"
| stats count by city
```

HTTP Status Codes by Time

To create this report, observe the following recommendations:

- **Panel Type:** Line Chart

- **Time Range:** All time

- **Search**:

```
index=transactions sourcetype=transaction_data
status=*
| timechart count span=1d by status
```

HTTP Methods by Time

To create this report, observe the following recommendations:

- **Panel Type:** Line Chart

- **Time Range:** All time

- **Search**:

```
index=transactions sourcetype=transaction_data
method=*

| timechart count span=1d by method
```

Validate the Dashboard

You can rearrange the panels to divide the dashboard as you want. Figure 6-9 shows our solution, with the following layout:

1. **Row 1**: The three Total Transaction requests

2. **Row 2**: The three Successful Transaction requests

3. **Row 3**: HTTP information

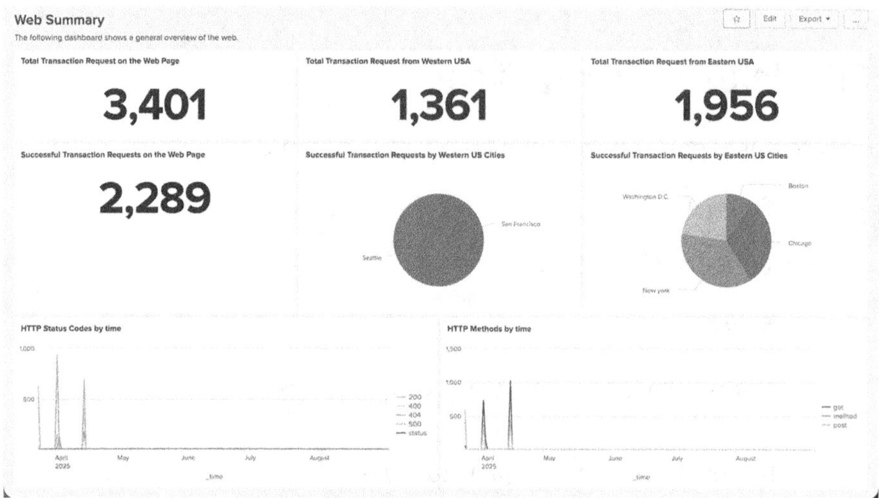

Figure 6-9. *Web summary dashboard*

You can check the code from your dashboard by going to it and clicking Edit ➤ Source. You will be able to find the Web summary dashboard code in GitHub: https://github.com/cmoreno94/splunk-certification-book/blob/main/web_summary.xml.

Dashboard Studio

In this section, you learn to create a dashboard using Dashboard Studio, using the same queries that we created in the previous section. You can reuse the same search string.

1. Go to the certification_book app ➤ Dashboards ➤ Create New Dashboard.

2. Add the following information:

   - **Dashboard Title**: Web Summary (Studio).

   - **Description**: The following dashboard shows a general overview of the web.

   - **Permissions**: Shared in App.

   - **Dashboard type**: Dashboard Studio.

   - **Layout type**: Grid.

Validate the configuration as shown in Figure 6-10.

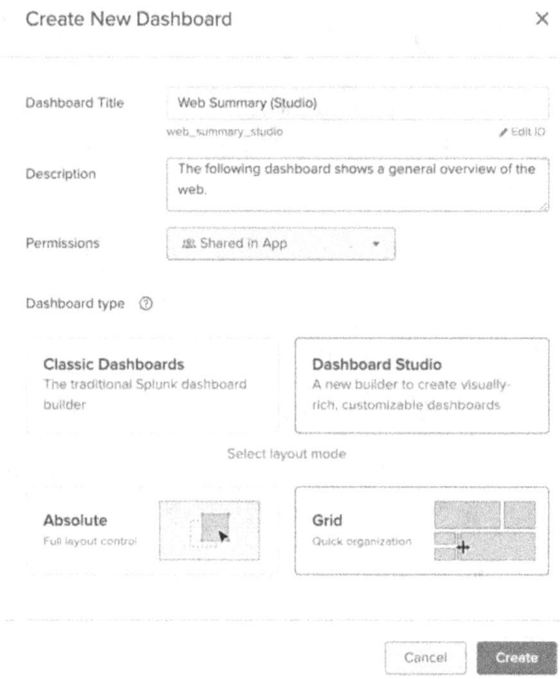

Figure 6-10. *Web summary: Dashboard Studio*

Note   You can have a copy of the classic dashboard created in the previous topic using "Clone in Dashboard Studio."

Once the dashboard is created, it's time to prepare the searches for the panels. We will pick an example, and you can replicate it along with the other queries discussed in the previous topic:

1. In the dashboard you have just created, click Data sources (database icon).

2. Click Create search, and complete the following information:

- **Data source name**: Total Transaction Request on the Web Page

- **SPL query**:

```
index=transactions sourcetype=transaction_data |
stats count
```

- **Time range**: Static/All time

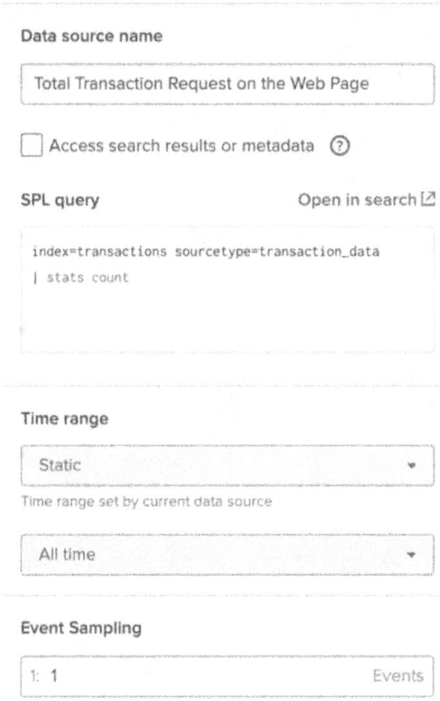

Figure 6-11. *Total Transaction Request on the Web Page data source*

Repeat the process for the rest of the searches. Once you have created the data sources, you will be ready to use them in the panels. Check the following steps to create the visualization for the Total Transaction Request on the Web Page data source:

1. Click visualizations, the icon with two bars, and select Single value.

2. Click Total Transaction Request on the Web Page data source, and automatically, the result of the data source will populate the visualization.

3. Repeat the process with the rest of the data sources.

4. Validate the final dashboard as shown in Figure 6-12.

Figure 6-12. *Web Summary (Studio) dashboard*

The next section of this chapter discusses dynamic form-based dashboards and their input fields.

Dynamic Form-Based Dashboards

Dynamic form-based dashboards allow users to filter data without moving to a new page. It enables you to add input fields, such as radio buttons, link lists, time modifiers, and drop-downs. This type of dashboard is used for troubleshooting, getting insights into data, and attaining business intelligence.

Type of Filters

Time range: It is a control element that allows the user to choose a time frame and filter data based on category and time.

Drop-down: It allows the user to choose only one of the available options and filter data based on that value. It can be a static list, a dynamic one, or a combination of both.

Multiselect: Similar to the drop-down, but the user will be able to select multiple options.

Text: It allows the user to enter specific text to filter the data sources in the dashboard.

Number: Same as text, but just with numbers.

Button: An Interactive object to configure actions in the dashboard, such as link, submit changes, etc.

We have prepared a dashboard with some filters; you can get it from `https://github.com/cmoreno94/splunk-certification-book/blob/main/web_summary_studio.json`. The following filters were added:

- **Time range**: It was applied to all the data sources. Now, the data sources will get the events based on the time defined in the time range filter.

- **Methods**: It is a drop-down filter with three options (All, GET and POST). It will filter the results of the first three panels.

- **HTTP code**: It is a multiselect filter, with one predefined option (All), and a list of options based on the results of a new data source, which gets the HTTP codes detected.

In Figure 6-13, you can check the result.

Figure 6-13. *Web Summary (Studio) with filters*

Summary

Splunk is a user-friendly platform for searching, processing, and monitoring big data. It encourages easier code and commands in a hierarchical arrangement of its elements and objects. This chapter furthered the discussion on the knowledge objects. It also talked about the knowledge managers, the people responsible for maintaining knowledge objects. You analyzed the dashboards and saw how they easily provide big data visualization.

The next chapter is a Splunk Core Certified User and Power User mock test.

You have covered 6% of Module 6 (Creating Reports and Dashboards) and part of the Splunk Basics, knowing more about the knowledge objects.

Multiple-Choice Questions

A.   In the Reassign Knowledge Objects page in Settings, you can view shared knowledge objects.

1. True

2. False

B.   Knowledge objects are stored in______. (Select all options that apply.)

1. macros.conf

2. tags.conf

3. eventtypes.conf

4. savedsearches.conf

5. props.conf

6. transforms.conf

C.   A knowledge manager's role includes (Select all options that apply.)

1. Granting users access to appropriate knowledge objects

2. Building data models for Pivot

3. Normalizing data

4. None of the above

D.   Which is a type of knowledge object? (Select all options that apply.)

1. Data interpretation (fields and field extraction)

2. Data classification (event types and transactions)

3. Data enrichment (lookups and workflow actions)

4. Data normalization (tags and aliases)

5. Data models

6. None of the above

E.   Select the filter elements of a dashboard.

1. Lookups

2. Time modifier

3. Text

4. Drop-down

5. Aliases

6. None of the above

Answers

- a. 1

- b. 1, 2, 3, 4, 5, 6

- c. 1, 2, 3

- d. 1, 2, 3, 4, 5

- e. 2, 3, 4

References

- https://help.splunk.com/en/splunk-enterprise/
  get-started/search-tutorial/10.0/part-7-
  creating-dashboards/create-dashboards-
  and-panels

- https://help.splunk.com/en/splunk-enterprise/
  developing-views-and-apps-for-splunk-web/10.0/
  getting-started/building-customizations-for-
  the-splunk-platform

Splunk User/Power User Exam Set

This chapter features multiple-choice questions (MCQs) that are useful in preparing for Splunk Core Certified User and Splunk Core Certified Power User certifications. In this section, you get a better idea of the kinds of questions that appear on these certification exams.

Questions

A. What are the main components of Splunk?

   1. Splunk forwarder, Splunk indexer, and Splunk search head

   2. Splunk indexer, Splunk heavy forwarder, and Splunk search head

   3. Splunk indexer, Splunk deployment manager, and Splunk forwarder

   4. Splunk heavy forwarder, Splunk deployment manager, and Splunk indexer

© Carlos Moreno Buitrago, Deep Mehta 2026

C. M. Buitrago and D. Mehta, *The Splunk Core User Study Companion*, Certification Study Companion Series, https://doi.org/10.1007/979-8-8688-2501-9_7

B. When you install Splunk on a stand-alone machine, which components are on it?

1. Input data

2. Parser

3. Indexer

4. All of the above

C. When you install the Splunk Enterprise free 60-day trial version, what is your daily index size?

1. 100 MB

2. 500 MB

3. 750 MB

4. 1000 MB

D. A _______ displays statistical trends over time.

1. Chart value

2. Time series

3. Stats value

4. Table value

E. The _____ command can display any series of data you want to plot.

1. dedup

2. stat

3. timechart

4. where

F. Which are filtering commands in Splunk?

1. where

2. search

3. eventstats

4. eval

G. A transaction is a grouping command.

1. True

2. False

H. Splunk extracts default fields when adding data.

1. True

2. False

I. Delimiters are used for ____ data.

1. Structured

2. Unstructured

J. Which default fields are metadata in Splunk?

1. Host, datatype, and sourcetype

2. Host, sourcetype, and source

3. 1 and 2

4. None of the above

K. When you run macros in Splunk, you need to rewrite the entire command.

1. True

2. False

L. Select the option that consists of lookup types.

1. CSV lookup

2. Geospatial lookup

3. KV Store lookup

4. External lookup

5. All of the above

M. Tags are used for ____ pairs.

1. Field/value

2. Key/value

3. 1 and 2

4. Field/key

N. As a knowledge manager, you are responsible for building data models that provide Pivot.

1. True

2. False

O. Data model acceleration doesn't use automatically generated summaries to speed up completion time in Pivot.

1. True

2. False

P. To create an alert in Splunk using .conf, you must
    edit ____.

    1. savedsearches.conf

    2. props.conf

    3. alert.conf

    4. transformation.conf

Q. You can access a tag in Splunk using ____. (Select all
    options that apply.)

    1. ::

    2. =

    3. *

    4. &

R. Field aliases appear in interesting fields if they
    appear at least ____.

    1. 10%

    2. 80%

    3. 40%

    4. 20%

S. Select all options that are data model dataset
    hierarchy types.

    1. Events

    2. Transactions

    3. Searches

    4. None of the above

Answers

A. 1

B. 4

C. 2

D. 2

E. 3

F. 1, 2

G. 1

H. 1

I. 1

J. 2

K. 2

L. 5

M. 3

N. 1

O. 2

P. 1

Q. 1, 2, 3

R. 4

S. 1, 2, 3

PART II

Splunk Data Administration and System Administration

Splunk Licenses, Indexes, and Role Management

This chapter covers the three pillars that govern cost, data, and access in Splunk. First, you'll understand licensing, how Splunk measures usage, how license pools/quotas are organized, and what happens when you exceed them. Next, you'll design indexes, the storage layer that controls retention, performance (hot/warm/cold), and data separation for search and security. Finally, you'll implement role-based access control (RBAC), building roles with the right capabilities and index permissions so users see only what they should and nothing they shouldn't. The following topics are covered in this chapter:

- Buckets in Splunk

- journal.gz, .tsidx, and Bloom filter

- Splunk licenses

- Managing Splunk licenses

- User management

© Carlos Moreno Buitrago, Deep Mehta 2026
C. M. Buitrago and D. Mehta, *The Splunk Core User Study Companion*, Certification Study
Companion Series, https://doi.org/10.1007/979-8-8688-2501-9_8

Buckets

Buckets are on-disk directories that store compressed raw data and TSIDX index files. There are three types of buckets:

- **Hot buckets**: After data is parsed into Splunk, it goes through a license manager, and the event is written in a hot bucket. Put simply, it is the bucket where data is written. Hot buckets are searchable. A hot bucket switches to a warm bucket when the maximum size or time is reached or Splunk is restarted. An index may have multiple hot buckets open at a time.

- **Warm buckets and cold buckets**: Warm buckets are searchable, but they cannot be written. Warm buckets are identified on "_time". Warm buckets are rolled to cold buckets when a particular index exceeds a size limit. After a particular time, cold buckets are rolled to freezing or archived, depending on the index policy.

- **Freezing**: Bucket that has aged out of an index per retention (time/size) and is removed from searchable storage. By default, it's **deleted**; optionally, it can be **archived** via a cold-to-frozen script and later **restored** by placing it in the thawedPath.

The indexes.conf file manages index policies such as data expiration and data thresholds. Splunk rolls a hot bucket to warm based on size (maxDataSize) and/or time span (maxHotSpanSecs), and it enforces overall retention per index using age (frozenTimePeriodInSecs) and/or total size (maxTotalDataSizeMB).

How Does a Bucket Work?

When you ingest data in a Splunk platform through monitor input, scripted input, or data upload, it is indexed. Splunk saves it in directories called *buckets*. According to data policy definitions, the first bucket is the hot bucket, where the data stays for some time. It is a read/write bucket.

A hot bucket rolls into a warm bucket in either case: The splunkd (a service that accesses, processes, and indexes streaming data and handles search requests) is restarted, the index size exceeds the limit, the timespan of the bucket is too large, the hot bucket has no receiving data in a while, or there is an increase in the bucket's metadata. Subsequently, it creates a new directory named "db_[newest_time]_[oldest_time]_ [ID]". It is named on "_time" to know the events' time range in a particular bucket.

Time in a warm bucket is expressed as an epoch. A warm bucket is read-only and rolls to the cold bucket in either case if the index limit is reached or too many of them are created. In this rollover from warm bucket to cold bucket, the entire bucket directory is copied to a cold bucket path. Ideally, the oldest index would be switched first.

Cold buckets allow you to read data and then either delete the data or roll it over to a frozen bucket, as defined in the policy. In the frozen bucket, the index cannot be read, and you cannot write data, only archive it. To read frozen data, restore it to the thawed path. The oldest bucket is deleted from the index when the index's maximum size is reached or the bucket age reaches the limit. A frozen path can be configured. This flow is represented in the diagram in Figure 8-1.

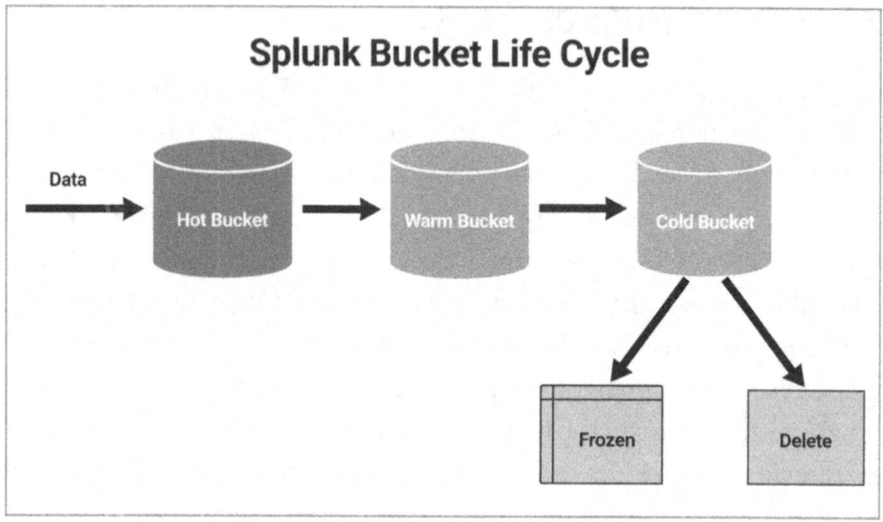

Figure 8-1. *Bucket life cycle*

The idea of having multiple bucket types (hot/warm/cold/frozen) is to control the data lifecycle and improve the usage of multiple storage tiers. Hot and warm buckets should use faster storage, such as SSD. Cold buckets can use slower storage, such as SAN/NAS.

How Search Is Performed in Buckets

When you search indexed data, Splunk returns results in reverse chronological order. Let's suppose that you want to search data in a recently added event. Splunk directly looks in hot buckets because they are the buckets that have been recently written to. So, it is easy to find recent results.

Similarly, if you search for the keyword *kite*, Splunk checks the hot bucket first. If the word is found, Splunk displays the timestamp of the event "kite" and the event's actual time coming to the bucket. However, it then looks for the event in the warm bucket in the reverse chronology of _time and, if not found, moves to the cold bucket, searching in the same manner. Therefore, in Splunk, the search mainly occurs by checking the recently added events through _time, digging into the bucket only when the keywords match. Henceforth, if a "kite" event is newly added and another "kite" event is in the warm or cold bucket, Splunk shows the event in the hot bucket first, followed by the event in the warm or cold bucket.

Let's now discuss the three commands used by the buckets to perform a search: journal.gz, .tsidx, and Bloom filters.

Understanding journal.gz, .tsidx, and Bloom Filters

To understand the workings of a search, you need to know the three components that buckets use to perform a search:

- **journal.gz**: When you ingest data in Splunk, it is saved in hot buckets. Inside hot buckets, journal.gz plays a crucial role in saving data. The raw data is compressed and divided into slices. A slice can be up to 128 KB of uncompressed data.

- **.tsidx** stands for *time-series index file*. Each event is given a unique ID within a bucket. An event is segmented into terms and saved in lexicographical (sorted alphabetically) order. Each term is tracked in its .tsidx, and a binary search can determine if the term is there or not. If the term is found, it can be scanned linearly to check whether data is there. A lexicon table points it to the address where the raw data is stored. Together, journal.gz and the .tsidx make a bucket.

- **Bloom filters** decrease an indexer's time to retrieve events from an index. A Bloom filter plays a crucial role in predicting whether an event is present in the index or not. Bloom filters run at the bucket level. For example, if you run a search command with the keyword *error*, a Bloom filter filters all buckets that don't contain that keyword in the lexicon table. Splunk Enterprise saves you time by searching tsidx files within a specified bucket where the search content is available.

Now you know about buckets and their components. But how does the Splunk search function work?

How Do Search Functions Work?

Suppose that you want to search for events with the keyword *kite*. Splunk first narrows buckets by the search time range, then uses a per-bucket Bloom filter to skip buckets that cannot contain the term. For buckets that pass, Splunk consults the tsidx to get seek offsets and retrieves the matching events directly from rawdata/journal.gz. Let's now move on to discuss Splunk licenses.

Splunk Licenses

When you input data in Splunk, the indexer indexes it and stores it on the disk. The Splunk license determines the data ingestion limit. Each Splunk Enterprise instance needs a license that specifies the rules on the amount of data it can index in a day. There are various types of Splunk licenses:

- **Splunk Enterprise License**

  The "standard" Splunk Enterprise license defines the licensed daily indexing volume and enabled Enterprise features.

- **No-Enforcement License**

  The standard Splunk Enterprise license has a maximum daily indexing volume for which you get a violation warning if it is exceeded. If you receive more than five warnings in a month, you violate your license. This may result in the Splunk search head being disabled. However, in no-enforcement, even if you violate the license, your search head won't be disabled.

- **Enterprise Trial License**

  The Enterprise Trial license allows maximum indexing of up to 500 MB/day. It expires after 60 days, and you are asked to switch to the standard Splunk Enterprise license or the Free license.

- **Sales Trial License**

  The Enterprise Trial license expires after 60 days and has an indexing capacity of 500 MB/day. If you have a pilot project that needs indexing capacity greater than 500 MB/day, you can directly contact the Splunk sales team to get a license.

- **Dev/Test License**

  Splunk provides access to its Dev/Test license to operate in a non-production environment with up to 10 GB daily.

- **Free License**

  The Free license in Splunk has an indexing capacity of up to 500 MB/day. With this license, you cannot perform distributed searches, TCP/HTTP forwarding, alerts, user management, LDAP, or scripted authentication.

- **Forwarder License**

  The Forwarder license allows forwarding of unlimited data. Unlike a free license, it enables authentication. You need not buy an extra license because it is included in Splunk.

Changing a License Group in Splunk

To move from trial to Enterprise license, follow these steps:

1. Go to Settings and select Licensing.

2. Select the License Group that you want to implement.

A sample screen image is shown in Figure 8-2.

Change license group

The type of license group determines what sorts of licenses can be used in the pools on this license server. ☑ Learn more

○ **Enterprise license**

This license adds support for multi-user and distributed deployments, alerting, role-based security, single sign-on, scheduled PDF delivery, and unlimited data volume.

There are no valid Splunk Enterprise licenses installed. You will be prompted to install a license if you choose this option.

○ **Forwarder license**

Use this group when configuring Splunk as a forwarder. ☑ Learn more

○ **Free license**

Use this group when you are running Splunk Free. This license has no authentication or user and role management, and has a 500MB/day daily indexing volume. ☑ Learn more

◉ **Enterprise Trial license**

This is your included download trial. IMPORTANT: If you switch to another license, you cannot return to the Trial. You must install an Enterprise license or switch to Splunk Free.

Cancel   Save

***Figure 8-2.** Splunk license group*

Management is the soul of the undisturbed working of any program or software. Let's look at how Splunk licenses are managed.

Managing Splunk Licenses

Splunk licenses are based on the maximum amount of data you can index per day. For example, the Splunk Enterprise Trial license group provides 500 MB/day licensing capacity. All the Splunk Enterprise instances need a license according to usage and the type of data being input. The following describes the types of data:

- **Event data** is the "raw_data" that the indexer inserts into the indexing pipeline.

- **Metrics data** counts up to 150 bytes each. It draws from the same license quota as event data.

- **Summary indexing** lets you run fast searches over large datasets by spreading out the cost of a computationally expensive report over time.

License metering occurs in the indexing phase. Splunk licenses manage the work in two categories:

- License manager
- License peer

License Managers and Peers

In Splunk, the license manager and the license peer work in synchronization. In your enterprise instance, there is one license manager instance, and all other instances are peer nodes. Cluster manager, indexer, deployers, deployment server, Monitoring Console, etc. All other instances are generally peer nodes managed by the license manager. The role of license manager is to manage the license usage of your Enterprise instance.

License Manager

The license manager in Splunk manages license access from a central location when there are multiple instances. The license manager allocates licensing capacity and manages the license usage for all instances. You simply set one Splunk instance as the license manager, and the remaining are its license peers. To configure an instance as a license manager using Splunk Web, refer to the following steps:

1. Go to Settings and select Licensing.

2. Designate the license server type as manager.

License Peer

When you configure one Splunk instance as the license manager, the remaining instances need to be license peers. However, if you have a single instance, that node acts as the license manager without a license peer. To configure an instance as a license peer, follow these steps:

1. Go to Settings and select Licensing.

2. Designate the license server type as peer.

3. Specify which master it should report to. Provide an IP address or hostname along with the Splunk management port (the default is 8089).

4. Click Save and restart the Splunk instance.

Figure 8-3 shows changing the license type.

Change master association

This server, **Deeps-MacBook-AL.local**, is currently acting as a master license server.

○ Designate this Splunk instance, **Deeps-MacBook-AL.local**, as the master license server

Choosing this option will:

- Point the local indexer at the local master license server
- Disconnect the local indexer from any remote license server

● Designate a different Splunk instance as the master license server

Choosing this option will:

- Deactivate the local master license server
- Point the local indexer at license server specified below
- Discontinue license services to remote indexers currently pointing to this server

Master license server URI

https://192.168.0.101:8089

For example: https://splunk_license_server:8089

Use https and specify the management port.

Cancel   Save

Figure 8-3. *License manager*

If you have a dedicated license manager in your Enterprise instance, the number of peers is the only factor that needs to be considered. You can have low CPU cores, memory, and disk size.

The next section deals with the process of adding a license in Splunk.

Adding a License in Splunk

License addition is very important for the long-term implementation of the Splunk software. Suppose you have an Enterprise Trial license and want to switch to a standard Splunk Enterprise license. Similarly, suppose you have an indexing license that allows up to 500 GB/day, but your usage is increasing, and you need more. You can add more licenses.

To add a license in Splunk, follow these steps:

1. Go to Settings and select Licensing.

2. Select Add License.

3. To install a license, upload a license file (license files end with .license).

4. Click Install.

Figure 8-4 shows how to add a new license in Splunk.

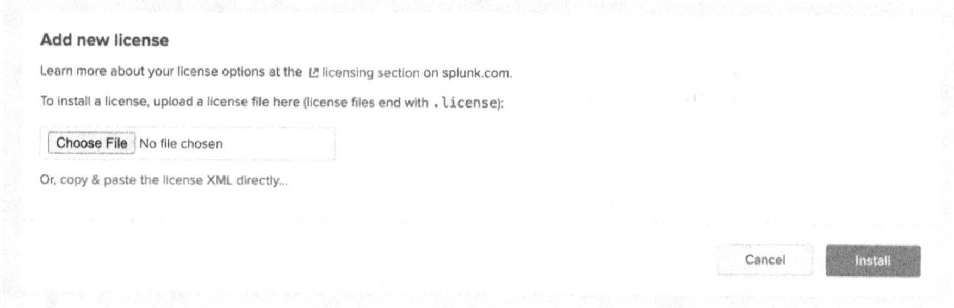

Figure 8-4. *Adding a license*

Splunk licenses are saved in $SPLUNK_HOME/etc/licenses.

License Pooling

License pooling allows licenses to be subdivided and assigned into groups of indexers. A license size is allocated to the master node, and the license manager manages the license pool and the peer nodes' access to the volume. If you have a Splunk standalone instance, it automatically becomes a licensed master, residing in the Splunk Enterprise stack. It has a default license pool named **auto_generated_pool_enterprise**.

Creating a License Pool

To create a license pool in Splunk Web, refer to the following:

1. Go to Settings and select Licensing.

2. Click Add Pool at the bottom of the page.

3. Specify a name for the pool.

4. Set the size allocation. Allocation is valuable. It can
    be the entire amount of indexing volume available
    in the stack; it depends on your priorities and
    requirements.

5. Specify the indexers that can access the pool. A
    particular indexer can have access to a particular
    pool, or a group of indexers may have access to a
    particular pool.

You cannot create a license pool in the Splunk Enterprise Trial license.

The index forms a very important part of the Splunk environment. It is the building block of the entire software. So, managing it efficiently is crucial. The following section discusses the management of the indexes in Splunk.

Managing Indexes in Splunk

In Splunk, the indexer processes data and saves it to the main index by default. But you can create and specify indexes for other data input. The index consists of files and directories, also called *buckets*. Buckets change per defined rules.

Along with the main index, Splunk also has preconfigured internal indexes.

- **main** is the default index where all incoming data is stored **unless** a different index is specified.

- **_internal** includes internal logs and metrics.

- **_audit** includes events related to auditing, user search history, and so forth.

- **_introspection** tracks system performance and usage resources.

- **summary** is the default index for the summary indexing system.

- **_thefishbucket** is Splunk's internal checkpoint database. It stores file CRCs and read offsets, so Splunk knows where it left off when monitoring inputs. While you can query it using _thefishbucket, it is not a standard event index and does not consume license volume.

The following are types of indexes:

- **Events indexes** impose a basic structure and allow all types of events to include metrics data.

- **Metrics indexes** are a highly structured format and impose low-latency demands related to event data. Putting metrics data into metrics indexes rather than events indexes leads to better performance and less index storage use. Metrics data is counted up to 150 bytes per event.

Creating an Index in Splunk

Creating an index segregates incoming data, which helps the Splunk admin bring the right team for the operation. For example, if I created an index named security to collect all events related to my organization's safety, it would help the company's security team. There are several steps to creating an index in Splunk.

Creating an Index Using Splunk Web

To create an index using the Splunk Web, take the following steps:

1. Go to Splunk Web and select Settings.

2. Go to Indexes and click New.

3. Enter the following information to create a
   new index:

   a. The name of your index.

   b. Enter HOME_PATH. The default path is $SPLUNK_
      DB/<index_name>/db.

   c. Enter COLD_PATH. The default path is $SPLUNK_
      DB/<index_name>/colddb.

   d. Enter THAWED_PATH. The default path is $SPLUNK_
      DB/<index_name>/thaweddb.

   e. Enable/Disable integrity check.

   f. Enter the maximum size of your index. The default is
      500,000 MB.

   g. You can also configure the size of each index bucket
      according to your needs or else select auto.

   h. If you want your data to be archived, enter the path in the
      frozen path.

   i. Select the app for which you want to create the index.

   j. For storage optimization, configure the .tsidx policy.

Figure 8-5 shows the screen for creating a new index in a Splunk
instance.

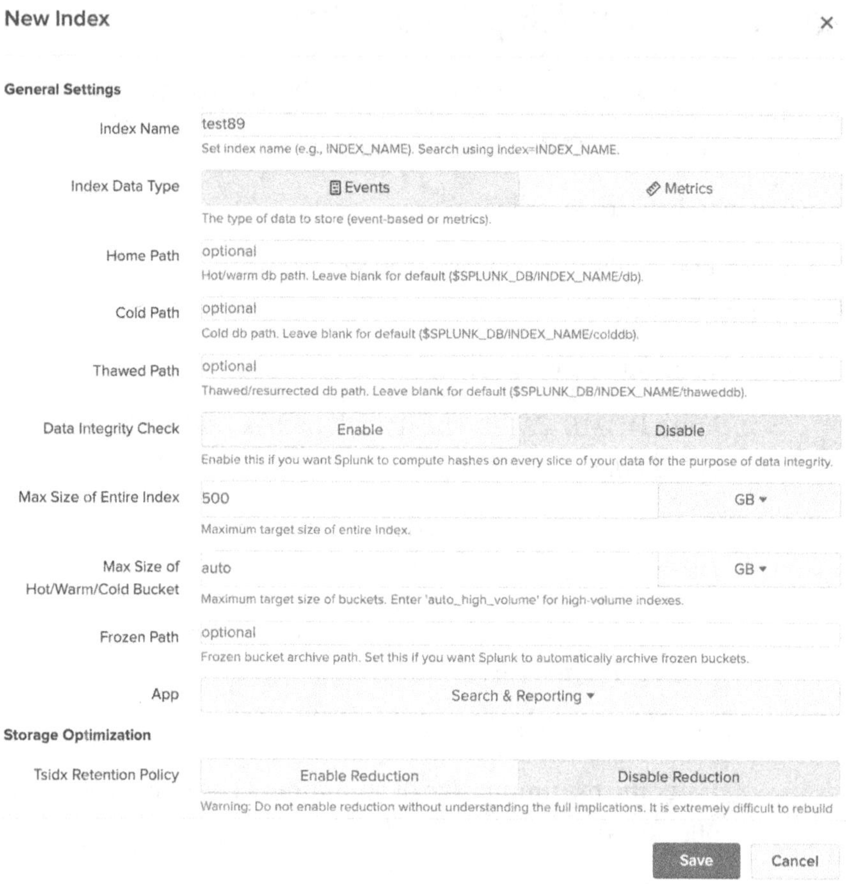

Figure 8-5. *Create index: test89*

Creating an Index Using a Splunk Configuration File

To create an index using a configuration file, edit the indexes.conf file located in $SPLUNK_HOME/etc/apps/<app>/local. $SPLUNK_HOME/

etc/system/local can be used as well, but the recommendation is to have a separate app to manage your indexes configuration files. Refer to the following code block:

```
[security_test]
homePath=<path>
coldPath=<path>
thawedPath=<path>
enableDataIntegrityControl=0|1
enableTsidxReduction=0|1
maxTotalDataSizeMB=<size>
```

1. In homePath, enter the path where you want to store your buckets in Splunk. The default path is $SPLUNK_DB/<index_name>/db.

2. In coldPath, enter the path where you want to store your cold buckets in Splunk. The default path is $SPLUNK_DB/<index_name>/colddb.

3. In the thawed path, enter the path where you want to store your archived buckets in Splunk. The default path is $SPLUNK_DB/<index_name>/thaweddb.

4. Enable Data Integrity Control to configure integrity checks in Splunk.

5. Enable Tsidx Reduction for storage optimization. Configure according to your needs.

6. In maxTotalDataSizeMB, configure the maximum size of your buckets. Please enter the size in MB.

Creating an Index Using Splunk CLI

To create an index using Splunk CLI, refer to the following steps:

```
splunk add index security_test <homepath> -coldPath <coldpath>
    -thawedPath <thawedpath> -enableDataIntegrityControl <0|1>
    -enableTsidxReduction <0|1> -maxTotalSizeMB<size>
```

1. In -homePath, enter the path where you want to
   store your buckets in Splunk. The default path is
   $SPLUNK_DB/<index_name>/db.

2. In -coldPath, enter the path where you want to store
   your cold buckets in Splunk. The default path is
   $SPLUNK_DB/<index_name>/colddb.

3. In -thawedPath, enter the path where you want to
   store your archived buckets in Splunk. The default
   path is $SPLUNK_DB/<index_name>/thaweddb.

4. In -enableDataIntegrityControl, configure for
   integrity checks in Splunk.

5. Configure -enableTsidxReduction, which covers
   storage optimization, according to your needs.

6. -maxTotalDataSizeMB configures the maximum size
   of your buckets. Please enter the size in MB.

You've now wrapped up this section. The last section of this chapter discusses user management and software privileges.

User Management

User management covers the privileges that a user has within Splunk. It determines the roles and working structure of the Splunk admin. There are various user management tasks that a Splunk admin needs to handle.

- Adding local users

- Defining role inheritance and role capabilities

- Role index search options

Adding a Local User

Adding a local user to Splunk is a simple process. To add a local user in Splunk (web), refer to these instructions:

1. Go to Settings and select Users.

2. Click New User.

3. Provide a username and password (required).

4. Provide the user's full name and email address (defaults to none).

5. Enter your time zone (defaults to search head time zone).

6. Enter the default app (defaults to home).

7. Enter the role (defaults to user).

Figure 8-6 is a screenshot of adding a native user to Splunk.

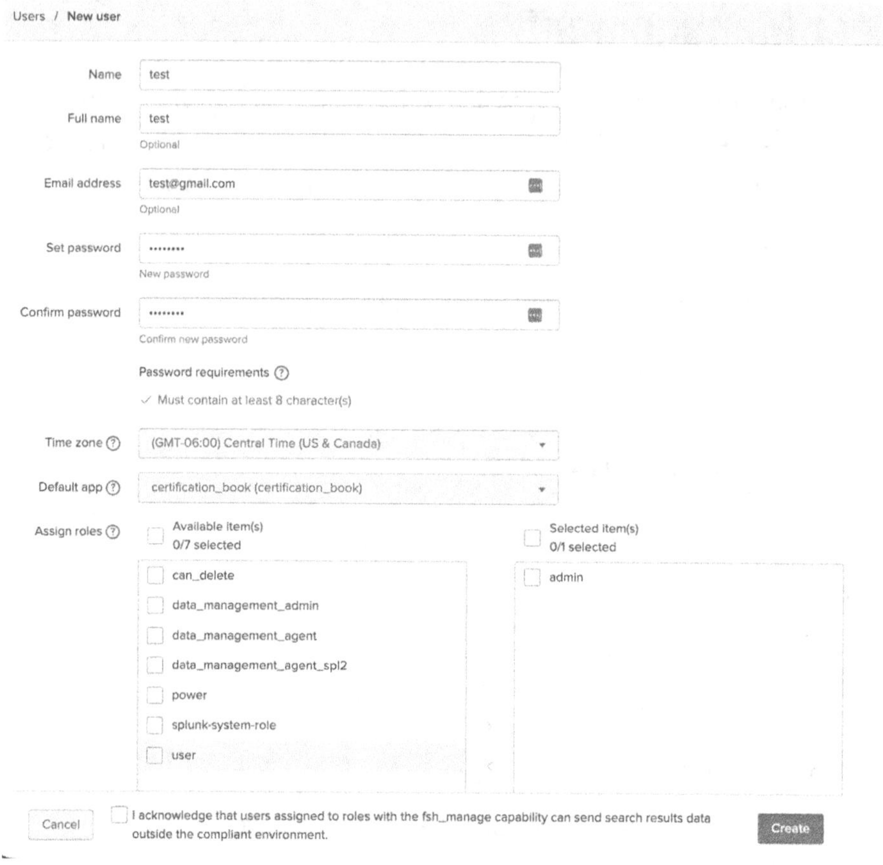

Figure 8-6. *Adding a local user*

For local Splunk authentication, hashed passwords are stored in $SPLUNK_HOME/etc/passwd.

Defining Role Inheritance and Role Capabilities

A role is a collection of permissions and capabilities that defines a user function in the Splunk platform. Users can have one or more roles. By default, Splunk Enterprise comes with the following roles predefined:

- The admin role has the most capabilities assigned to it.

- The power role can edit all shared objects (saved searches, etc.) and alerts, tag events, and other similar tasks.

- The user role can create and edit its own saved searches, run searches, edit its own preferences, create and edit event types, and other similar tasks.

- The can_delete role allows the user to delete by keyword. This capability is necessary when using the delete search operator.

A new user can inherit capabilities and index access from other roles. The following steps define role inheritance:

1. Go to Settings and select Roles.

2. Select the role you want to update.

3. Go to Inheritance, and select whichever roles you want to inherit.

Figure 8-7 shows adding inheritance in Splunk.

Edit Role admin

Name * ⑦   admin

Enter a name of the role

Inheritance   Capabilities   Indexes   Restrictions   Resources

Select roles from which this role inherits capabilities and indexes. Inherited capabilities and indexes cannot be deselected. If you select multiple roles, this role inherits capabilities and indexes from all of them.

■	Role name ▼
☐	can_delete
☐	data_management_admin
☐	data_management_agent
☐	data_management_agent_spl2
☑	power
☐	splunk-system-role
☑	user

Figure 8-7. *Defining role inheritance to Splunk users*

The following steps add role capabilities in Splunk:

1. Go to Settings and select Roles.

2. Select whichever role you want to update.

3. Go to Capabilities, and select Assign or Remove Capabilities.

Figure 8-8 is a screenshot of adding role capabilities to Splunk.

Edit Role admin

Name * ⑦   admin

Enter a name of the role

Inheritance   **Capabilities**   Indexes   Restrictions   Resources

Select capabilities for this role.

☑ Capability name ▼	Inheritance
☑ accelerate_datamodel	native
☑ accelerate_search	inherited
☑ admin_all_objects	native
☑ apps_backup	native
☑ apps_restore	native
☑ capture_ingest_events	native
☑ change_audit	native
☑ change_authentication	native
☑ change_own_password	inherited
☑ create_external_lookup	native
☐ delete_by_keyword	
☑ dispatch_rest_to_indexers	native
☑ edit_authentication_extensions	native
☑ edit_bookmarks_mc	native
☐ edit_cam_queue	
☑ edit_certificates	native

Cancel   Save Role

Figure 8-8. *Defining role capabilities to Splunk roles*

With edit roles and edit user capabilities, users can get/avoid
capabilities, which elevates the security and compliance of your Splunk
Enterprise environment.

Additionally, you can choose the indexes that the role is allowed to query.
In Role Index Access Options, if an index is not selected, users cannot
access it.

1.   Go to Settings and select Roles.

2.   Select whichever role you want to update.

3. Go to Index Searched by default, and assign or remove an index.

Figure 8-9 shows adding index access to Splunk users.

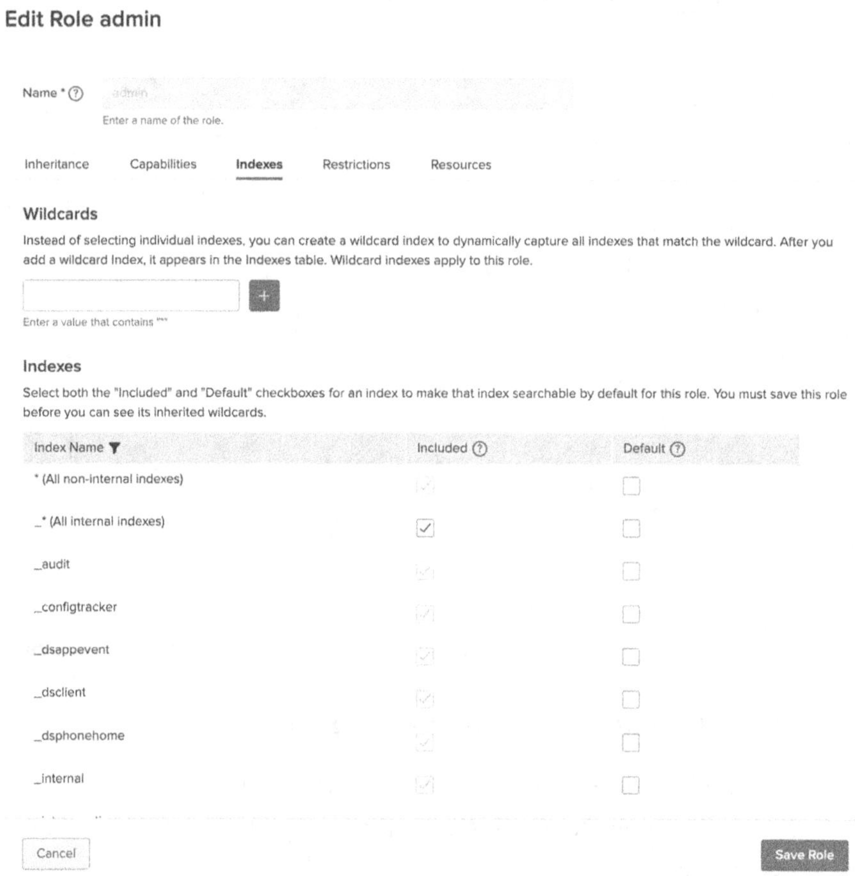

Figure 8-9. *Defining index access to Splunk users*

You have now come to the end of the chapter. Commend yourself! You can test your knowledge by practicing with the MCQs at the end of the chapter.

Summary

With the completion of Chapter 8, you have covered a big portion of the Splunk software basics. From tags and keywords to admins and licenses, you have learned most of Splunk's major concepts. This chapter dealt with Splunk licenses, license pooling, managing indexes, buckets, journal.gz, .tsidx, Bloom filters, and user management. Therefore, you are now focusing on the executive elements and the administrative and management elements as well.

According to the Splunk Enterprise Certified Admin exam blueprint, you covered 5% with Module 2 (License Management), 10% with Module 4 (Splunk Indexes), and 5% with Module 5 (Splunk User Management).

The next chapter covers Splunk forwarder and clustering.

Multiple-Choice Questions

A. Select all the main components of an index in Splunk.

1. journal.gz

2. .tsidx

3. Bloom filter

4. None of the above

B. What is the maximum capacity of the Splunk Enterprise Trial License for indexing per day?

1. 500 MB

2. 700 MB

3. 1 GB

4. 100 MB

C. Which are the preconfigured internal indexes?
(Select all options that apply.)

1. main

2. _internal

3. _audit

4. _introspection

5. summary

6. test

7. None of the above

D. You can create a Splunk index using the following
options. (Select all options that apply.)

1. Splunk CLI

2. indexes.conf

3. Splunk Web

4. inputs.conf

5. transforms.conf

E. In which phase of the index time process does
license metering occur?

1. Input phase

2. Parsing phase

3. Indexing phase

4. Licensing phase

F. Local user accounts created in Splunk store passwords in which file?

1. $ SPLUNK HCME/etc/users/authentication.conf

2. $ SPLUNK_HOME/etc/passwd

3. $ SFLUNK_KCME/etc/authentication

4. $ SFLUNK_KOME/etc/passwd

G. Metrics data is counted against the Enterprise license at a fixed 120 bytes per event.

1. True

2. False

H. A Splunk event in _internal index is counted against licensing.

1. True

2. False

Answers

A. 1, 2, 3

B. 1

C. 1, 2, 3, 4, 5

D. 1, 2, 3

E. 3

F. 2

G. 2

H. 2

References

- https://help.splunk.com/en/splunk-enterprise/
  administer/manage-indexers-and-indexer-
  clusters/10.0/manage-index-storage/how-the-
  indexer-stores-indexes

- https://docs.splunk.com/Splexicon:Bloomfilter

- https://conf.splunk.com/files/2016/slides/
  behind-the-magnifying-glass-how-search-
  works.pdf

- https://docs.splunk.com/Splexicon:Tsidxfile

Machine Data Using Splunk Forwarder and Clustering

When handling a large amount of data from sources, data collection becomes necessary for the systematic procedure of the functions. Firstly, this chapter discusses the Splunk forwarder and its types. Secondly, you will also see how to ingest data from the Splunk forwarder into the Splunk indexer, demonstrating the power of machine data. Thirdly, you explore the process of clustering so that you can implement clusters into data collection. Finally, you will see LDAP and SAML used for authentication and credential management in Splunk.

The following topics are covered in this chapter:

- Splunk universal forwarder

- Splunk light and heavy forwarder

- Forwarder management (deployment server)

- Splunk indexer clusters

- Splunk Lightweight Directory Access Protocol (LDAP)

- Splunk Security Assertion Markup Language (SAML)

© Carlos Moreno Buitrago, Deep Mehta 2026
C. M. Buitrago and D. Mehta, *The Splunk Core User Study Companion*, Certification Study Companion Series, https://doi.org/10.1007/979-8-8688-2501-9_9

Splunk Universal Forwarder

The Splunk universal forwarder collects data from the source and forwards
it to the receivers using minimum hardware resources. It is lighter because
it doesn't have a web interface, but the cost of installation needs to be
considered, depending on how much data it should read/forward. The
universal forwarder includes only those components that are necessary to
forward data to Splunk instances. It is managed by editing configuration
files, editing the deployment server.

Configuring Splunk Indexer to Listen to Data for Universal Forwarder

The following explains the process of forwarding data into the Splunk
indexer through Splunk Web:

1. Go to the Splunk indexer instance web interface.

2. Go to Settings and select Forwarding and Receiving.

3. Go to Configure Receiving and Add New.

4. Enter **9997** in **Listen on this port**. (Any port
   you wish, but if you allocate a different port, in
   a forwarder instance, you need to allow it in the
   firewall rules.) Refer to Figure 9-1.

Figure 9-1. Configure listening on indexer

Do the following steps to enable port listening in the indexer using Splunk CLI:

1. In Linux, go to the Splunk indexer, and using the terminal, run:

   $SPLUNK_HOME/bin/splunk enable listen <port number>.

2. In Windows, go to the Splunk indexer, and using Command Prompt, run:

   %SPLUNK_HOME%\bin\splunk enable listen <port number>.

Any OS indexer platform is supported when sending logs from a universal forwarder. Let's now look at the configuration used in the Windows Splunk forwarder.

Configuring Windows Splunk Forwarder

There are two ways to configure the Windows Splunk universal forwarder:

- By using the command line

- By executing the .msi files

Windows can execute .msi files, or you can use the command line to install Splunk Universal Forwarder. Download Splunk universal forwarder for Windows from `www.splunk.com/en_us/download/universal-forwarder.html`.

Splunk Universal Forwarder Using Windows

To configure the Splunk universal forwarder using a command line, refer to the following steps:

1. Go to Command Prompt, and go to %SPLUNK_
    HOME%\bin.

2. To start, stop, and restart Splunk, type **.\splunk
    start**, **.\splunk stop**, and **.\splunk restart**.

3. Forward data to the indexer by running the
    following command:

    ```
    .\splunk add forward-server <IP or
    Hostname>:<Port>
    ```

4. Go to SPLUNK_HOME\etc\system\local\outputs.
    conf to check whether the universal forwarder is
    forwarding data or not. The following is an output
    example:

    ```
    [tcpout]
    defaultGroup = default-autolb-group
    [tcpout:default-autolb-group]
    disabled = false
    server = 192.168.1.20:9997
    ```

Splunk Universal Forwarder Using .msi

To configure Splunk universal forwarder using .msi, download the Splunk
.msi file from www.splunk.com/en_us/download/universal-forwarder.
html. To configure the universal forwarder, refer to the following
instructions (see Figure 9-2):

1. Check the box to accept the license agreement, and
    then click Next.

2. Enter the username and the password.

3. Enter the Deployment Server hostname or IP and
    the port. Skip this step for now.

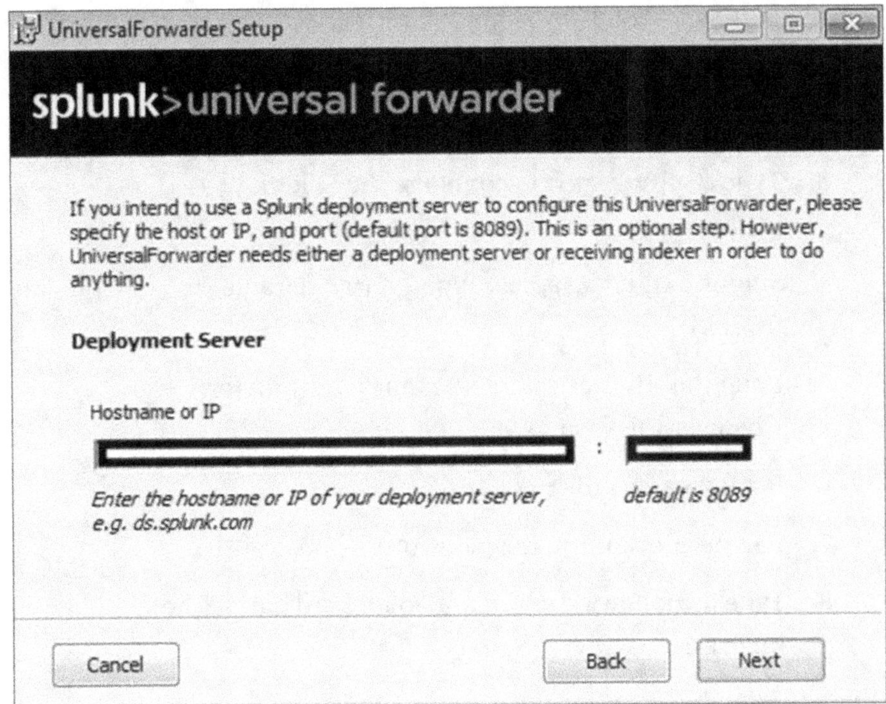

Figure 9-2. *Configure universal forwarder to indexer*

Let's now look at configuring the Linux Splunk forwarder.

Configuring Linux Splunk Forwarder

Linux Splunk forwarder can be downloaded through a command-
line widget, .rpm file, .tgz file, or .deb file. To install Splunk universal
forwarder through an .rpm package file, .tgz package file, or a .deb package
file, download from `www.splunk.com/en_us/download/universal-`
`forwarder.html`.

Let's now configure the Splunk forwarder by using the terminal in Linux.

1. In the terminal, go to $SPLUNK_HOME/bin.

2. To start, stop, or restart Splunk, type **./splunk start**, **./splunk stop**, or **./splunk restart**.

3. Type **./splunk start** to configure the Splunk instance.

4. Splunk asks if you agree. Type "**y**". (y = yes and n = no.)

5. Enter the administrator username as **admin** (or whatever you wish).

6. Enter a password.

7. Retype and confirm the password.

8. Type the following command to forward data to the indexer:

```
./splunk add forward-server <IP or Hostname>:<Port>
```

9. Go to `$SPLUNK_HOME/etc/system/local/outputs.conf` to check whether the universal forwarder is forwarding data or not. You should see output similar to the following:

```
[tcpout]
defaultGroup = default-autolb-group
[tcpout:default-autolb-group]
disabled = false
server = 192.168.1.20:9997
```

Next, let's discuss the two types of forwarders:

- Light forwarder

- Heavy forwarder

Splunk's Forwarders

You have learned that a forwarder collects data from sources, forwards it to an indexer, and configures it using different methods. Now you learn about the heavy forwarder and other old options that can appear in the exams.

Splunk Heavy Forwarder

A heavy forwarder has a smaller footprint than a Splunk indexer. It does almost all the jobs of the indexer except for distributed searching. It parses data before forwarding and routes data based on criteria such as the source or type of event. A heavy forwarder also has the capability to index data locally and send it to a different indexer, but it is not a common case.

Configuring Heavy Forwarder

To configure a heavy forwarder, observe the following steps:

1. Install Splunk Enterprise on the machine that you want to configure the heavy forwarder.

2. Click Settings and go to Forwarding and Receiving.

3. Select Configure Forwarding and click Add New.

4. Enter the indexer hostname or IP and the port (please enable the listening port on the indexer).

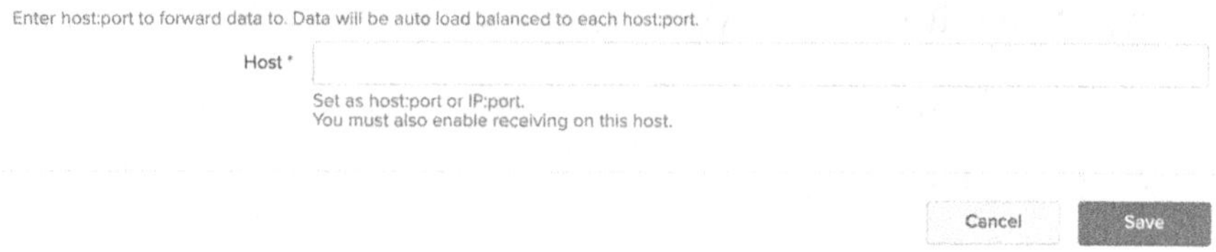

Figure 9-3.   Configure Splunk heavy forwarder to indexer

5.   Restart the Splunk instance.

Configuring Heavy Forwarder to Index and Forwarding Data from a Universal Forwarder

The Heavy Forwarder can collect data from APIs (using Splunk Technology Add-ons), TCP/UDP, Splunk universal forwarder, etc. To configure a heavy forwarder to index and forward the data, observe the following steps:

1.   Go to the Splunk Enterprise instance web interface.

2.   Click Settings and go to Forwarding and Receiving.

3.   Go to Configure Receiving and click New.

4.   In **Listen to this port**, type **9997**. (Use any port you wish.) Refer to Figure 9-4.

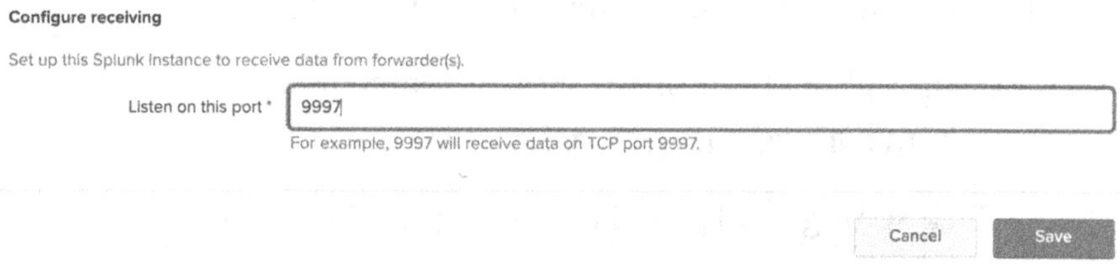

Figure 9-4.   Configure receiving on Splunk heavy forwarder

Now, the heavy forwarder (HF) is ready to get data from the Universal Forwarders. The heavy forwarder is the type of Splunk forwarder that can parse data before forwarding it to the indexers. It acts as a central point of collection before sending the data to your indexers. Imagine that you have a decent number of universal forwarders (UF) across multiple regions, and the Splunk environment is just in one of the regions; if you want to reach out to the Splunk indexers with all your UFs, you should request a lot of firewall requests to allow the traffic. Set an HF in each region, and they will take care of sending the data to Splunk indexers.

Splunk Light Forwarder

The light forwarder was deprecated from Splunk Enterprise version 6.0, being replaced by the universal forwarder. So, configuration of the light forwarder is only possible on versions prior to Splunk Enterprise 6.0.

Let's now focus on the deployment server.

Deployment Server

This section discusses the management of the forwarder and its configuration.

The Splunk deployment server provides a centralized node to manage the deployment clients. The deployment server role comprises checking the system's status, monitoring deployment activity, and configuring and uninstalling the apps in your deployment clients. It provides an interface to create server classes, mapping clients' deployment to the deployment of apps alongside. Its main objective is to check the system's status, monitor deployment activity, and configure and uninstall the apps in your environment. It allows you to modify the server classes and push an update. Simply put, agent management provides a centralized node to manage most of the Splunk domain.

There are three main tabs in agent management.

- **Applications:** This tab displays a list of the apps that are deployed and shows the app status. You can push app updates from inside forwarder management and edit the apps' properties.

- **Server classes:** This tab displays a list of classes that are deployed and shows the server-class status. You can create new server classes and edit the existing ones.

- **Forwarders:** This tab displays a list of the deployed clients and their status. You can restrict views of the app through this tab.

The server class in the agent management interface determines the app collection that will be deployed on the forwarders.

In the next section, you will learn the various processes and platforms available to configure agent management.

Configuring Agent Management

The following steps set up agent management in a Splunk environment.

The Agent Management screen resembles the one shown in Figure 9-5.

Figure 9-5. *Configuring Agent Management*

1. Go to $SPLUNK_HOME/etc/deployment-apps/.

2. Add the certification app to the deployment-server folder (copy it from $SPLUNK_HOME/etc/apps/ certification_book).

3. Go to Settings in Splunk Web, and then go to Agent Management.

Figure 9-6 is a screenshot of the configured Agent Management page.

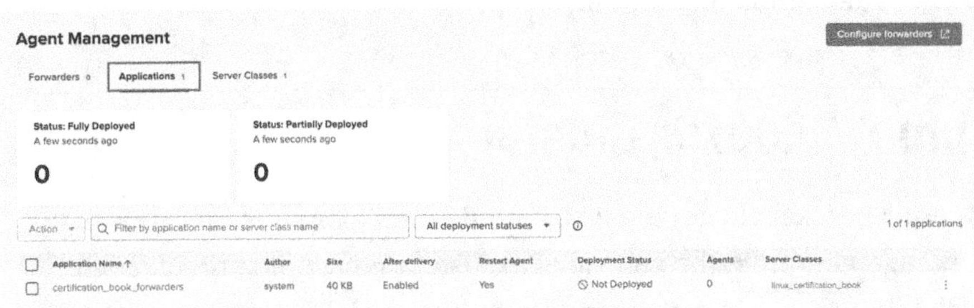

Figure 9-6. *Applications page*

Configuring the Forwarder Management Client

The forwarder can be a universal forwarder or a heavy forwarder. To configure the client for forwarder management, refer to the following rules:

1. Using a Linux terminal, run to $SPLUNK_HOME/ bin/splunk set deploy-poll (IP or hostname):(mgmt_port_number).

2. In Windows Command Prompt, go to SPLUNK_ HOME%\bin\splunk set deploy-poll (IP or hostname):(mgmt_port_number).

After setting deploy-poll, go to $SPLUNK_HOME/etc/system/local/ deploymentclient.conf. You find the deploymentclient.conf similar to the following:

```
[target-broker:deploymentServer]
targetUri= 192.168.1.20:8089
```

This sums up the discussion on the forwarder and its various elements. The next section discusses indexer clusters, which is another important aspect of data implementation.

Splunk Indexer Clusters

A Splunk indexer cluster is a distributed group of Splunk indexers that work together under the control of a cluster manager to provide high availability, data replication, and horizontal scalability for indexed data. Instead of sending data to a single indexer, data is written to one peer indexer and then automatically replicated to other peers in the cluster, so that multiple searchable copies of each bucket exist. This replication model protects against data loss if an indexer fails and allows searches to run in parallel across all indexers in the cluster for higher performance. In short, an indexer cluster is how Splunk ensures resilience, consistency, and scale for your data at the storage and search layer.

You can achieve the following points with index replication:

- Data availability

- Data fidelity

- Disaster tolerance

- Improved search performance

Configuring Indexer Clusters

The minimum indexer cluster architecture in Splunk includes two nodes, but it won't take all the benefits of a cluster. The recommendation is to have a minimum of three indexers or use SmartStore, which, instead of keeping all index buckets on local disks, the indexer keeps only the active, recently searched data in a local cache and offloads the rest of the bucket data to a remote object store (see Figure 9-7 for minimum index cluster).

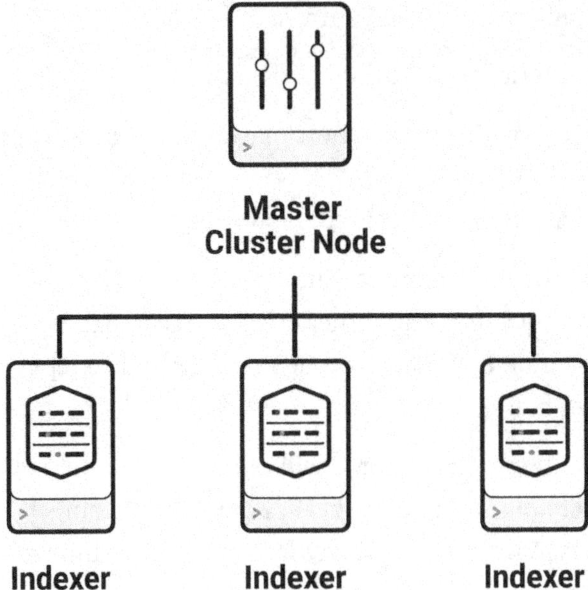

Figure 9-7. *Minimum recommended indexer cluster architecture*

- The **master node** manages the cluster. It keeps track of the replicating copies located at the peer nodes (slave nodes) and helps the search head find the cluster members. It also helps maintain automatic failover from one indexer to the next.

- The **peer node** receives and indexes incoming data from forwarders. A peer node can index its incoming data and simultaneously store copies of data from other nodes.

Creating an Indexer Cluster Using Splunk Web

To create an indexer cluster using Splunk Web, refer to the following steps:

1. In Splunk Web, go to Settings.

2. Select Indexer Clustering and enable it.

This is where you distinguish how master nodes and peer nodes are created.

To create a master node, enter the following:

1. Enter the replication factor. It acts as a backup of the buckets without the metadata, so a higher replication factor protects against loss of data if peer nodes fail.

2. Enter the search factor. It will replicate the searchable copies of a bucket across the cluster nodes. A higher search factor speeds up the time to search and recover lost data at the cost of disk space.

3. Create a security key. This key authenticates communication between the master and the peers.

4. Name your cluster. (This step is optional.)

Figure 9-8 shows an example of configuring a manager node.

Manager Node Configuration   ✕

Replication Factor

> 3

The number of copies of raw data that you want the cluster to maintain. A higher replication factor protects against loss of data if peer nodes fail.

Search Factor

> 2

The number of searchable copies of data the cluster maintains. A higher search factor speeds up the time to recover lost data at the cost of disk space. Must be less than or equal to Replication Factor.

Security Key

> •••••

This key authenticates communication between the manager and the peers and search heads.

Cluster Label

> dev

Name your cluster using this field. This label is also used to identify this cluster in the Monitoring Console.

Back   Enable Manager Node

Figure 9-8. *Configuring a manager node via Splunk Web*

Once the required details are filled in, click Enable Manager Node to create the node.

To configure a peer node, enter the following information:

1. Enter the manager node URI (Ip:mgmt_port/host_
   name:mgmt_port).

2. Enter the peer replication port. The port peer nodes
   use to stream data to each other.

3.   Use the security key created in the manager node.
     This key authenticates communication between the
     master and the peers.

4.   Click Enable peer node.

Figure 9-9 shows an example of configuring a peer node.

Peer node configuration   ✕

Manager URI   https://192.168.1.21:8089

                     E.g. https://10.152.31.202:8089

Peer replication port   8080

                     The port peer nodes use to stream data to each other (Eg:
                     8080).

Security key   •••••

                     This key authenticates communication between the manager
                     and the peers and search heads.

Back   Enable peer node

Figure 9-9. *Configuring a peer node: Splunk Web*

Creating an Indexer Cluster Using a Splunk .conf File

To configure a Splunk Manager Node from the .conf file, the manager
creates or edits server.conf in $SPLUNK_HOME/etc/system/local or
by deploying a specific app in $SPLUNK_HOME/etc/apps/<my_custom_
mc_app>.

The syntax for configuring a manager node using a .conf file is shown in the following code block:

```
[clustering]
mode = manager
replication_factor = <number>
search_factor = <number>
pass4SymmKey = <String>
```

- cluster_label = <String>

- Replication _factor protects against loss of data if peer nodes fail.

- search_factor speeds up the time to recover lost data at the cost of disk space.

- pass4SymmKey authenticates communication between the master and the peers.

- cluster_label is the name of the cluster.

- To configure a peer node, the indexer creates or edits server.conf located in $SPLUNK_HOME/etc/system/ local or by deploying a specific app in $SPLUNK_HOME/ etc/apps/<my_custom_indexer_app>.

The syntax for configuring a peer node using a .conf file is shown in the following code block:

```
[replication_port://<port>]
[clustering]
manager_uri = https://<manager_node>:<mgmt_port>
mode = peer
pass4SymmKey = <string>
```

- master_uri is the address of the manager node.

- pass4SymmKey authenticates communication between the master and the peers.

Creating an Indexer Cluster Using Splunk CLI

To configure an indexer cluster using Splunk CLI for a manager node, refer to the following steps.

1. Using Windows CMD or a Linux terminal, go to `$SPLUNK_HOME/bin`.

2. To start, stop, and restart Splunk, type **.\splunk start**, **.\splunk stop**, and **.\splunk restart**.

The syntax for configuring a master node using Splunk CLI is shown in the following code block:

```
splunk edit cluster-config -mode manager -replication_factor
<number> -search_factor <number> -secret <string> -cluster_
label <string>
splunk restart
```

The following describes the steps for a peer node:

1. Using Windows CMD or a Linux terminal, go to `$SPLUNK_HOME/bin`.

2. To start, stop, and restart Splunk, type **.\splunk start**, **.\splunk stop**, and **.\splunk restart**.

The syntax for configuring a peer node using Splunk CLI is shown in the following code block:

```
splunk edit cluster-config -mode peer -manager_uri
https://<ip>:<mgmt_port>,<host_name>:<mgmt_port> -replication_port
<port> -secret <string> -cluster_label <string>
splunk restart
```

You have learned about the Splunk forwarder, forwarder management, and indexer clusters. Together, they are responsible for the collection, allocation, and gathering of data from sources. Let's now move on to Splunk LDAP, which supervises authentication in the Splunk domain.

Splunk Lightweight Directory Access Protocol (LDAP)

Splunk LDAP plays a critical role in the authentication of the Splunk Web. It maintains user credentials like user IDs and passwords and information like hostname, bind DN, and user base DN. When you configure LDAP in Splunk Web, it handles authentication. Each time the user logs in, Splunk rechecks the LDAP server.

Creating an LDAP Strategy

To create an LDAP strategy, follow these steps:

1. Click Settings in the Splunk home page, and go to Authentication Methods.

2. Click Configure Splunk to use LDAP, and go to New LDAP. In the New LDAP, enter the following details:

   a. LDAP strategy name.

   b. The hostname of your LDAP server (ensure that the Splunk server can identify the hostname).

   c. The port that Splunk Enterprise uses to connect to the LDAP server defaults to port 636 (by default, LDAP servers listen on TCP port 389 and LDAPS (LDAP with SSL)).

3. Check if SSL is enabled. This is an optional step.

4. Provide the bind DN. (It defines the domain name of an administrator.)

5. Enter and confirm the bind DN password for the binding user.

6. Specify the user base DN (Splunk Enterprise uses this attribute to locate user information).
   The following is other information that may need to be addressed:

   - Userbase filter (recommended to return only applicable users; e.g., (department=IT, Management, Accounts, etc.))

   - User name

   - Real name attribute

   - Mail address

   - Group mapping attribute (a user attribute that group entries use to define their members)

   - Group base DN (the location of the user groups in LDAP)

   - Static group search filter for the object class you want to filter your static groups on

   - Group name attribute (a group entry attribute whose value stores the group name)

   - Static member attribute (a group attribute whose values are the group's members)

   For nested groups, check Nested groups. Enter the Dynamic group search filter to retrieve dynamic groups; if it exists, then enter the Dynamic member attribute.

In Advanced Settings, you can limit the search request time, search request size limit, and networking socket timeout.

7.   Click Save and exit.

Figures 9-10 and 9-11 show the process of LDAP strategy creation.

LDAP strategy name *

Enter a unique name for this strategy.

LDAP connection settings

Host *

Your Splunk server must be able to resolve this host.

Port

The LDAP server port defaults to 389 if you are not using SSL, or 636 if SSL is enabled.

☐ SSL enabled
You must also have SSL enabled on your LDAP server.

Bind DN

This is the distinguished name used to bind to the LDAP server. This is typically the DN of an administrator with access to all LDAP users you wish to add to Splunk. However, you can leave this blank if anonymous bind is sufficient.

Bind DN Password

Enter the password for your Bind DN user.

Confirm password

User settings

User base DN *

The location of your LDAP users, specified by the DN of your user subtree. If necessary, you can specify several DNs separated by semicolons.

User base filter

The LDAP search filter used to filter users. Highly recommended if you have a large amount of user entries under your user base DN. For example, "(objectClass=person)"

User name attribute *

The user attribute that contains the username. Note that this attribute's value should be case insensitive. Set to "uid" for most configurations. In Active Directory (AD), this should be set to "sAMAccountName".

Real name attribute *

The user attribute that contains a human readable name. This is typically "cn" (common name) or "displayName".

Email attribute

The user attribute that contains the user's email address. This is typically "mail".

Group mapping attribute

The user attribute that group entries use to define their members. If your LDAP groups use distinguished names for membership you can leave this field blank.

Figure 9-10. *Creating LDAP strategy*

Group settings

*Group base DN

The location of your LDAP groups, specified by the DN of your group subtree. If necessary, you can specify several DNs separated by semicolons.

Static group search filter

The LDAP search filter used to retrieve static groups. Highly recommended if you have a large amount of group entries under your group base DN. For example, '(department=IT)'

*Group name attribute

The group attribute that contains the group name. A typical value for this is 'cn'.

*Static member attribute

The group attribute whose values are the group's members. Typical values are 'member' or 'memberUid'. Groups list user members with values of groupMappingAttribute, as specified above.

Nested groups

Controls whether Splunk will expand nested groups using the 'memberof' extension. Only check this if you have nested groups and the 'memberof' extension on your LDAP server.

Dynamic group settings

Dynamic member attribute

The dynamic group attribute that contains the LDAP URL used to find members. This setting is required to configure dynamic groups. A typical value is 'memberURL'.

Dynamic group search filter

The LDAP search filter used to retrieve dynamic groups (optional). For example, '(objectclass=groupOfURLs)'

Advanced settings

Cancel   Save

Figure 9-11. *Creating LDAP strategy*

Mapping LDAP Group to Splunk Roles

Splunk Enterprise authenticates and maps the various roles via the LDAP server.

To map the LDAP group to Splunk roles, follow these instructions:

1. Select LDAP.

2. Configure Splunk to LDAP and map groups.

3. Select Map Groups in the Actions column for a specific strategy.

4. Click Group Name.

5. Map a role to a group, and click the arrow to a role in the Available Roles list. This moves the group into the Selected Roles list, which helps multiple roles to be mapped in the group (see Figure 9-12).

6. Click Save.

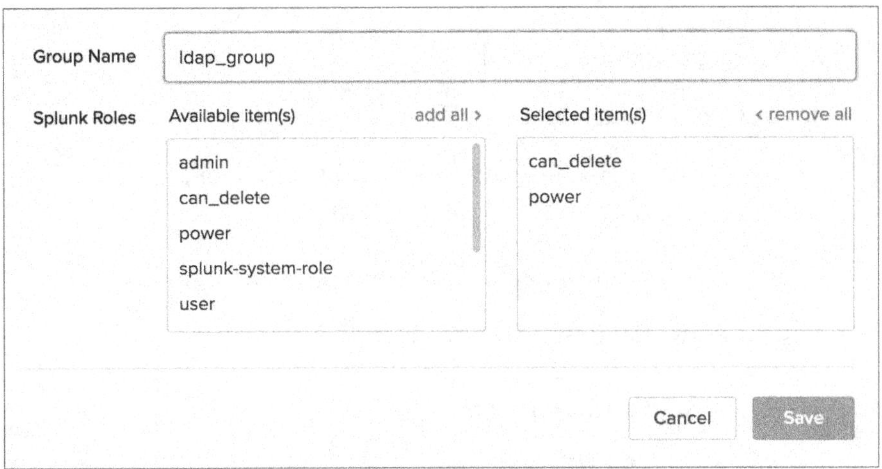

Figure 9-12. *Mapping LDAP strategy: Splunk roles*

That summarizes the Splunk LDAP analysis, its creation, mapping, and its importance in the Splunk domain. Let's move to Splunk SAML.

Splunk Security Assertion Markup Language (SAML)

Splunk SAML (Security Assertion Markup Language) is an identity provider (IdP) that maintains users' credentials and handles authentication. When you configure a Splunk instance to use the SAML authentication system, you authorize groups on your IdP and enable their login by mapping them to Splunk user roles. SAML supports authentication for SSO.

Splunk SAML and LDAP can also be configured using authentication. conf, located in $SPLUNK_HOME/etc/system/local.

Configuring Splunk SAML

To configure Splunk SAML, refer to the following steps:

1. Click Settings and go to Authentication Methods.

2. Click Configure Splunk to use SAML, and create a new SAML.

3. Copy the metadata file directly.

4. In General Settings, provide the following data:

    - Single sign-on URL (automatically populated when you enter the copy metadata file).

    - IdPs Certificate Path (it can be a directory or a file).

    - Entity ID (it is the entity's identity).

5. Select **Sign auth request**.

6. Select **Sign SAML response**.

7. If you use PingIdentity as your IdP, specify the attribute query URL, the sign attribute query request, and the sign attribute query response.

8. In Advance Settings, configure information like attribute alias role, attribute alias role name, and attribute alias mail.

9. Click Save.

Figure 9-13 illustrates the process of Splunk SAML configuration.

SAML Configuration   ×

Configure SAML for Splunk. Learn More

Download the SPMetadata from Splunk and add it to your SAML environment to connect to Splunk.

SP Metadata File   Download File

Import identity provider (IdP) metadata by browsing to an XML file, or copy and paste the information into the Metadata Contents text box.

Metadata XML File   Select File

Metadata Contents

Apply

General Settings

Single Sign On (SSO) URL *

Single Log Out (SLO) URL   optional

IdP certificate path *   optional

Leave blank if you plan to self-certify or when `SPLUNK_HOME/etc/auth/idp.pem`.

IdP certificate chain *

Replace Certificates *   ☑

Issuer *

Entity ID *

Sign AuthnRequest   ☑

Verify SAML response *   ☑

▶ **Attribute Query Requests**

▶ **Authentication Extensions**

▶ **Alias**

Advanced Settings

NameID Format *   ▾

Fully qualified domain name or IP of the load balancer *   optional

Redirect port - load balancer listener port *   optional

Redirect to URL after logout *   optional

SSO Accepting *   HTTP Post   HTTP Redirect

SLO Binding *   HTTP Post   HTTP Redirect

Cancel   Save

Figure 9-13. *Configuring SAML*

Let's look at mapping SAML.

Map SAML to User Roles

To map the SAML to Splunk roles, refer to the following instructions:

1. Go to Access Controls, and select Authentication Method.

2. Select SAML, and go to Configure Splunk to SAML and map groups.

3. On the SAML Groups page, click New Group, or click Edit to modify the existing group.

4. Provide a name for the group.

5. Determine the roles you want to assign to this group by moving the desired roles from the left column to the right column.

6. Click Save.

Figure 9-14 is an illustration of the SAML mapping method.

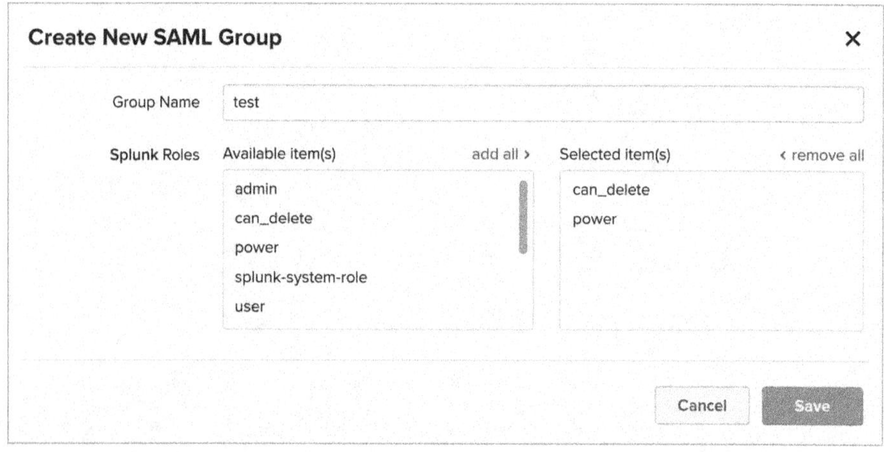

Figure 9-14. *Mapping SAML: Splunk roles*

This sums up the working of the Splunk SAML.

You have come to the end of this chapter. You should congratulate yourself on the successfully learning of Splunk data management. This chapter proves important when the need to sort or simplify data arises and takes you a step further in Splunk administration.

Summary

This chapter covered Splunk universal forwarder, the heavy and light forwarders, Splunk indexer clusters and their configuration, Splunk LDAP, and Splunk SAML. These topics administer the gathering, assembling, and authentication of the data in a Splunk environment. Data collection often includes remote sources and large-scale processing. Instance forwarders and indexer clusters ensure that all the data is assigned to the right buckets and sorted into the correct nodes. Splunk LDAP and SAML ensure data input is authenticated and the users' credentials are authorized. You learned implementation, configuration, and role mapping in step-by-step discussions.

The next chapter discusses how to fetch data using monitor and network inputs, extracting data using scripted and Windows input, pulling data using agentless input, and data sanitizing using fine-tuned input.

Practice what you learned in this chapter and test your knowledge with the following questions.

Multiple-Choice Questions

    A.   Which forwarder type can parse data before forwarding?

        1. Universal forwarder

    2. Heaviest forwarder

    3. Light forwarder

    4. Heavy forwarder

B. Which Splunk indexer operating system platform is supported when sending logs from a Windows universal forwarder?

    1. Any OS platform

    2. Linux platform only

    3. MAC OS platform only

    4. None of the above

C. Which Splunk component requires a forwarder license?

    1. Search head

    2. Heavy forwarder

    3. Heaviest forwarder

    4. Universal forwarder

D. Which of the following authentication types requires configuration in Splunk?

    1. ADFS

    2. LDAP

    3. SAML

    4. RADIUS

E. How often does Splunk refresh the LDAP server for group/role mapping?

1. Every five minutes

2. Each time a user logs in

3. Each time Splunk is restarted

4. Varies based on the groupMappingRefreshInterval in authentication.conf

F. Which of the following statements describes a deployment server?

1. An Enterprise license.

2. Is responsible for managing apps for forwarders.

3. Once used, it is the only way to manage forwarders.

4. Automatically restarts the host OS running the forwarder.

G. What is the minimum recommended number of indexers in a cluster?

1. 1

2. 2

3. 3

4. 5

H. How can you configure the Splunk index in Splunk? (Select all that apply.)

1. Cluster manager

2. Splunk CLI

3. indexes.conf

4. Splunk Web

Answers

A. 4

B. 1

C. 2

D. 2, 3

E. 4

F. 2

G. 3

H. 1, 2, 3, 4

References

- `https://help.splunk.com/en/splunk-enterprise/administer/manage-users-and-security/10.0/use-ldap-as-an-authentication-scheme/set-up-user-authentication-with-ldap`

- `https://help.splunk.com/en/splunk-enterprise/administer/manage-users-and-security/10.0/use-saml-as-an-authentication-scheme-for-single-sign-on/configure-single-sign-on-with-saml`

- `https://help.splunk.com/en/splunk-enterprise/administer/manage-users-and-security/10.0/use-saml-as-an-authentication-scheme-for-single-sign-on/configure-sso-with-pingidentity-as-your-saml-identity-provider`

- `https://help.splunk.com/en/splunk-cloud-platform/forward-and-process-data/universal-forwarder-manual/10.0/about-the-universal-forwarder/about-the-universal-forwarder`

- `https://docs.splunk.com/Splexicon:Heavy forwarder`

- `https://help.splunk.com/en/splunk-enterprise/forward-and-process-data/forwarding-and-receiving-data/10.0/deploy-heavy-and-light-forwarders/deploy-a-light-forwarder`

Advanced Data Input in Splunk

This chapter builds on forwarder advanced inputs and dives deeper into how Splunk collects data. You'll learn practical ways to bring in files, streams, scripts, and remote sources. The following topics are covered in this chapter:

- Additional forwarding options

- Fetching data using monitor input

- Scripted input

- Network input

- Pulling data using agentless input

In the first section, you analyze additional forwarding options. Through these options, you can compress a data feed before sending data to a Splunk indexer and acknowledge the indexed data to prevent data loss. These additional forwarding options secure a data feed so that you can maintain data integrity in Splunk.

- **Compress the data feed:** Compressing a data feed in Splunk can be done by editing outputs.conf in the forwarding environment and inputs.conf on the receiving side. If you don't compress the data feed, the

© Carlos Moreno Buitrago, Deep Mehta 2026

C. M. Buitrago and D. Mehta, *The Splunk Core User Study Companion*, Certification Study Companion Series, https://doi.org/10.1007/979-8-8688-2501-9_10

forwarding environment directly sends the raw events. However, this option compresses the raw events before transferring.

- **Indexer acknowledgment:** Indexer acknowledgment in Splunk is done by editing outputs.conf in forwarding environmental inputs.conf on the receiving side. Using indexer acknowledgment, the forwarder resends any data not acknowledged as received by the indexer.

- **Securing the feed:** Securing the feed (SSL) in Splunk is done through outputs.conf and server.conf. To add SSL on the receiving side, you can edit inputs.conf. Once you have secured your feed using the default SSL certificate or with your own, you can maintain data integrity. Additionally, the compress feature will be automatically applied.

- **Forwarding environment queue size:** The queue size in Splunk is modified through the outputs.conf maximum queue size, which is 500 KB by default. It is the maximum amount of data the forwarder has as a buffer when it cannot reach out to the server.

Let's now discuss their commands and syntax.

Compress the Data Feed

Splunk compresses raw data from the forwarding environment and sends it to the receiving side. On the receiving side, you need to decompress the data feed to get raw data and increase CPU usage. compressed=true lets the forwarder environment compress the data before it sends data to receivers.

To compress the data feed in the forwarder environment, edit outputs. conf as shown in the following code block:

```
[tcpout:splunk_indexer]
server=<ip/hostname>:<receiving_port>
compressed=true
```

To compress the data feed on the receiving side, edit inputs.conf using the following code block command:

```
[splunktcp:<receiving_port>]
compressed=true
```

With this, you have completed learning the additional forwarding options and the compression of raw data syntax. Next, let's discuss indexer acknowledgement.

Indexer Acknowledgment

Indexer acknowledgment helps prevent data loss when the forwarding environment is forwarding data to the receiving side. The forwarding environment resends the data if the Splunk indexer does not acknowledge it. When you use indexer acknowledgement in the forwarder, it is recommended to increase the output queue to 3× so that the requirement for a larger space is met.

To allow the indexer to acknowledge the forwarder environment, edit outputs.conf as shown in the following code block:

```
[tcpout:splunk_indexer]
server=<ip-hostname>:<receiving_port>
useACK=true
```

To allow the indexer to acknowledge the forwarder environment, edit the inputs.conf, as shown in the following code block:

```
[splunktcp:<receiving_port>]
useACK=true
```

When indexer acknowledgment (useACK=true) is enabled on a forwarder and maxQueueSize is left at its default auto, Splunk automatically sizes the forwarder's queues for reliability: the output queue is set to 7 MB, and the wait queue size is automatically set to three times the output queue size, 21 MB. The wait queue stores copies of recently sent data blocks until the indexer confirms (ACKs) that those blocks were safely written to disk, and its size is always three times the output queue size.

Securing the Feed

Securing the feed (SSL certificate) is important so that no one can easily send data to the indexer or manipulate the events. To add SSL on the receiving side, you need to edit inputs.conf, and to add it on the sender side, you need to edit outputs.conf and server.conf. Securing the feed using SSL increases CPU usage and automatically compresses the data feed.

To secure the feed in the forwarder environment, edit outputs.conf as shown in the following code block:

```
[tcpout:splunk_indexer]
server=<ip_hostname>:<receiving_port>
sslPassword=<password>
clientCert=<path_server.pem_file>
```

To secure the feed in the forwarder environment in Linux, edit the server.conf as shown in the following code block:

```
[sslConfig]
sslRootCAPath=<path_cacert.pem_file>
```

To secure the feed in the receiving environment, edit inputs.conf as follows:

```
[splunktcp:<receiving_port>]
[ssl]
sslPassword=<password>
serverCert=<path for server.pem file>
requireClientCert=false
```

To secure the feed in the receiving environment, edit server.conf (not in Windows) as follows:

```
[sslConfig]
sslRootCAPath=<path_cacert.pem_file>
```

Let's discuss queue sizes in the following section.

Queue Size

In a Splunk forwarding environment, queue sizes play a crucial role in managing data flow and indexer acknowledgment. While you can't directly configure the wait queue size, you can adjust the output queue size through the outputs.conf file. By default, Splunk defines these limits automatically:

- When indexer acknowledgment (useACK=false) is disabled, the output queue size is 500 KB.

- When indexer acknowledgment (useACK=true) is enabled, the output queue increases to 7 MB, and the wait queue expands to 21 MB.

These defaults help ensure that data is reliably sent from the forwarder to the indexer without overloading network or system resources.

The recommended setting for most environments is to use maxQueueSize = auto. To set the queue size to auto in the forwarder environment, edit outputs.conf as shown in the following code block:

```
[tcpout:splunk_indexer]
server=<ip_hostname>:<receiving_port>
maxQueueSize=auto
```

Monitor Input

A monitor input is the act of watching a file, directory, script output, or network ports for new data. There are two types of wildcards that you can add in inputs.conf to monitor files/directories, explained in Table 10-1.

Table 10-1. *Wildcards for monitor inputs*

Wildcard	Description
...	The ellipsis wildcard recurses through directories and subdirectories until a match.
*	The asterisk wildcard matches files/directories anything in the specific directory path.

Examples:

[monitor:///var/log/.../app.log] ➤ It walks through any subfolder under /var/log until it finds files named exactly app.log.

[monitor:///var/log/nginx/*.log] ➤ It matches all files ending in .log in that folder only. It won't go into subdirectories.

[monitor:///opt/myapp/*/logs/*.log] ➤ The first * matches exactly one folder name under /opt/myapp/, then looks for logs/*.log inside each of those.

Monitor Files

A monitor input defines a specific file as a data source; it performs the following tasks on a data source:

- The current data of the file is read and sent to be indexed.

- The file is continuously monitored for new content.

- Splunk tracks the file status, even after restarting a Splunk instance. It automatically points to the location from where the Splunk instance stopped tracking the file.

A file monitor in Splunk supports the following data types:

- Plain text data files

- Structured text files

- Multiline logs

- The ability to read compressed files

Monitor Directories

A monitor input defines a specific directory as a data source. Monitor input performs the following tasks in a data source:

- Splunk recursively travels through a directory structure.

- All types of files in the directory are included for monitoring.

In a monitor input, you can use blacklists and whitelists to monitor files and directories:

- **Whitelist**: A filtering rule to include one or more members from a set. You use whitelist rules to tell a forwarder which files to consume when monitoring directories. You can define a whitelist in inputs.conf.

- **Blacklist**: A filtering rule to exclude one or more members from a set. You use blacklist rules to tell a forwarder which files not to consume when monitoring directories. You can define a blacklist in inputs.conf.

Monitor files support CSV, XML, and JSON formats.

Monitor Files and Directories Using Splunk Web

To monitor files and directories using Splunk Web, refer to the following instructions. In this section, you monitor the LKDC-setup.log file, in which clients want files continuously monitored. You can monitor whatever file you want.

1. Go to Settings and go to Add Data.

2. Go to monitor and select Files & Directories.

3. Click Browse. Navigate to files. Select the file. Refer to Figure 10-1 for reference. (I selected the KDC-Setup.log file to be monitored.)

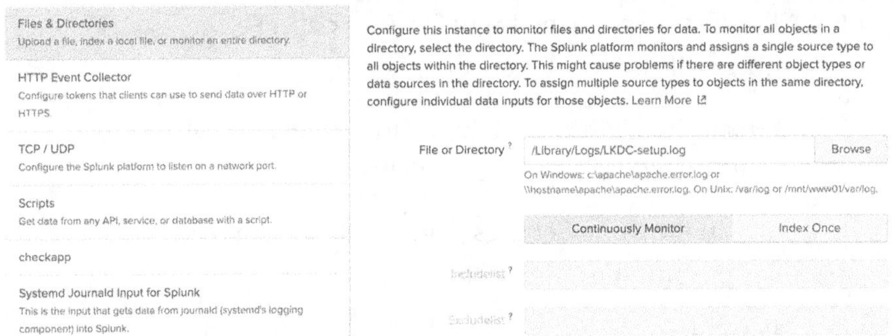

Figure 10-1. *Monitor files using Splunk Web*

4. Click Continuously Monitor or Index Once, depending on your requirements. We selected the Continuous Monitor option.

5. Click Next to go to the Set Source Type page (select the source type based on your data). Refer to Figure 10-2.

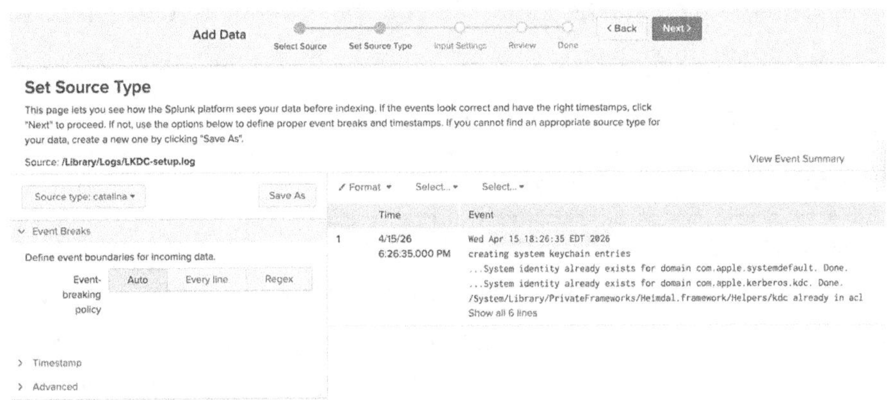

Figure 10-2. *Set sourcetype*

6. Click Next to go to the Input Settings page. Select App: certification_book.

7. In the Index Data Type field, verify that the default events index is selected, and click Save.

8. Back on the Input Settings page, the test should be displayed for the index value.

9. Click Submit.

Monitor File and Directory Using inputs.conf

You can directly monitor files or directories in Splunk using inputs.conf. You can directly monitor files by writing inputs.conf in the Splunk instance or deploy them using the deployment server. Refer to the following steps to configure inputs.conf to monitor files in Splunk.

Using the terminal, go to $SPLUNK_HOME/etc/apps/certification_book/local/inputs.conf (if it doesn't exist, create a new file named inputs.conf), and define the stanza. Refer to Table 10-2 for a description of the attributes of the stanza.

Table 10-2. *Monitor file using inputs.conf*

Attribute	Value
disabled=<truelfalse>	When disabled= input script won't run and when disabled= input script invokes.
whitelist	A filtering rule to include one or more members from a set. You use whitelist rules to tell a forwarder which files to consume while monitoring directories.
blacklist	A filtering rule to exclude one or more members from a set. You use blacklist rules to tell a forwarder which files not to consume while monitoring directories.

(continued)

Table 10-2. (*continued*)

Attribute	Value
host=<hostname>	Splunk uses the hostname during parsing and indexing.
index=<Index name>	Splunk sets an index to save an event.
sourcetype=<sourcetype name>	Source type in Splunk is used for formatting data during parsing and indexing.
source=<sourcename>	Source name in Splunk is an input file path where data originates.

As an example, if clients want to continuously monitor the LKDC-setup.log file, you can directly write a stanza using inputs.conf. (Refer to the following solution.)

1. Go to `$SPLUNK_HOME/etc/apps/certification_book/local/inputs.conf`. Create a stanza as follows:

```
[monitor:///Library/Logs/LKDC-setup.log]
disabled=false
sourcetype=lkdc
```

2. Save the changes.

3. Restart the Splunk instance.

Next, let's discuss the scripted input.

Scripted Input

Splunk scripted input can accept data from scripts. You can use scripted input to get data to Splunk Enterprise from a third-party application, application programming interface (API), remote data interface, and so on. Scripted input onboards data to Splunk Enterprise. Scripted input can be added to Splunk instances through Splunk Web or by editing inputs.conf. When you configure a Splunk environment to run a scripted input, it clears any environment variables that can affect the script's operation.

Splunk executes scripts from `$SPLUNK_HOME/etc/apps/<app_name>/ bin`, `$SPLUNK_HOME/bin/scripts` or `$SPLUNK_HOME/etc/system/bin`.

Scripted Input Using Splunk Web

In scripted input, you write a script to get the total size of RAM, the size of one block of RAM, total used blocks, and the interval every 60 seconds. After every 60 seconds, the local instance using scripted input sends data to the Splunk instance. The following is reference code in Python:

```
import sys
print("Total size",sys.getsizeof({}))
print("size of 1 block",sys.getsizeof([]) )
print("used block",sys.getsizeof(set()) )
```

1. Go to Settings and go to Add Data.

2. Go to Monitor and select Scripts.

3. Select the script path where your script is placed. (Normally, place your script in the app's bin folder.)

4. Select the script from the drop-down.

5. In the interval, select seconds or a cron schedule, whichever method you want.

6. Provide parameters based on whether the interval
    you select is seconds or a cron schedule. Refer to
    Figure 10-3 for reference.

Configure this instance to execute a script or command and to capture its output as event data.
Scripted inputs are useful when the data that you want to index is not available in a file to monitor.
Learn More

Script Path	$SPLUNK_HOME/etc/apps/certification_book/bin ▾
Script Name	1.py ▾
Command ?	$SPLUNK_HOME/etc/apps/certification_book/bin/1.py
Interval Input ?	In Seconds ▾
Interval ?	60.0
Source name override ?	optional

Figure 10-3. *Adding 1.py script*

7. Click Next.

8. In the Input Settings page:

    a. Select the sourcetype based on your data. You can also create
        a new source type. (The source type is one of the default fields
        that the Splunk platform assigns to all incoming data. It tells
        the Splunk platform what kind of data you have so that it can
        intelligently format the data during indexing.)

    b. In-App, select the certification_book app.

    c. In sourcetype, select Test18.

    d. In Index, select the index named *Test*.

    e. Click Next.

    f. Click Review.

Click Finish. (Refer to Figure 10-4.)

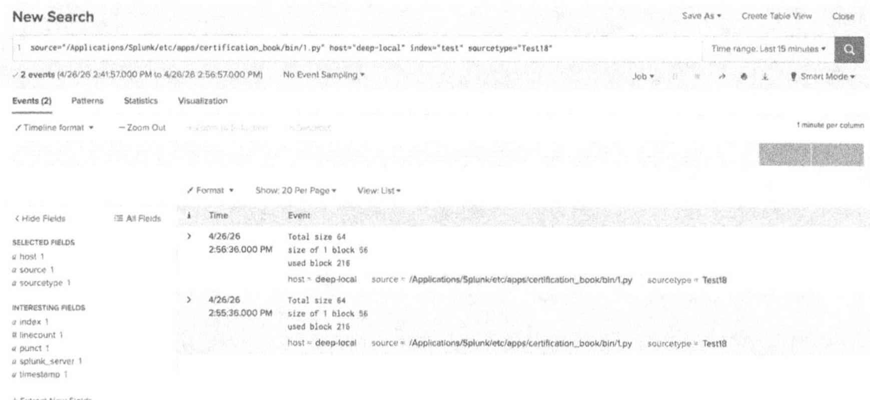

Figure 10-4. *Scripted input: 1.py*

Scripted Input Using inputs.conf file

Let's add scripted input using the inputs.conf file. The script needed for scripted input can reside in any one of the following locations:

- $SPLUNK_HOME/etc/system/bin

- $SPLUNK_HOME/etc/apps/<your_apps>/bin

- $SPLUNK_HOME/bin/scripts

You can directly ingest scripted input in Splunk by modifying inputs. conf. The various parameters for modifying inputs.conf are shown in Table 10-3.

Table 10-3. *Scripted input inputs.conf*

Attribute	Value
disabled=<true|false>	When disabled= input script won't run, and when disabled= input script invokes.
interval=<number>|cron schedule	You can specify the time interval. Set interval=<value> After that interval, the input script runs again. Use interval=<cron> when you want to specify a cron schedule, so the script does not execute on startup, but rather at the times that the cron schedule defines.
host=<hostname>	Splunk uses the hostname during parsing and indexing.
index=<Index name>	Splunk sets an index to save an event.
sourcetype=<sourcetype name>	Source type in Splunk is used for formatting data during parsing and indexing.
source=<sourcename>	Source name in Splunk is an input where data originates.

In scripted input, you write a script to get the total size of RAM, the size of one block of RAM, total used blocks, and the interval every 60 seconds. After every 60 seconds, the local instance using scripted input sends data to the Splunk instance. You write a script stanza in inputs.conf to monitor scripted input.

1. Copy 1.py code in the certification_book application bin folder.

2. Go to $SPLUNK_HOME/etc/apps/certification_book/local/inputs.conf. Create a stanza as follows:

```
[script://$SPLUNK_HOME/etc/apps/certification_
book/bin/1.py]
```

```
disabled=false
sourcetype=Test18
interval=60
index=Test
```

3. Save the changes. Restart the Splunk instance.

After completing the Scripted input, you now move on to the next section—the network inputs.

Network Input

Splunk Enterprise can accept input on any TCP (Transmission Control Protocol) or UDP (User Datagram Protocol) port. Splunk can consume any data on these ports. Syslog is a good example of network-based data. TCP is the recommended protocol for sending data in Splunk.

With a TCP input, you need to specify whether the port should accept connections from all hosts or specific hosts. The host value can be IP, DNS, or custom (user-defined label). To set up network input in Splunk, the network protocol and port number need to be specified.

Add Network Input Using Splunk Web

The following explains how to add a network input in Splunk:

1. In Splunk, click Settings, go to Add Data, and click Monitor.

2. In Select Source, click TCP/UDP and configure. Provide the port on which you want to listen to data. Provide a source name, and you can configure it to accept data from a particular forwarder. Refer to Figure 10-5 for help.

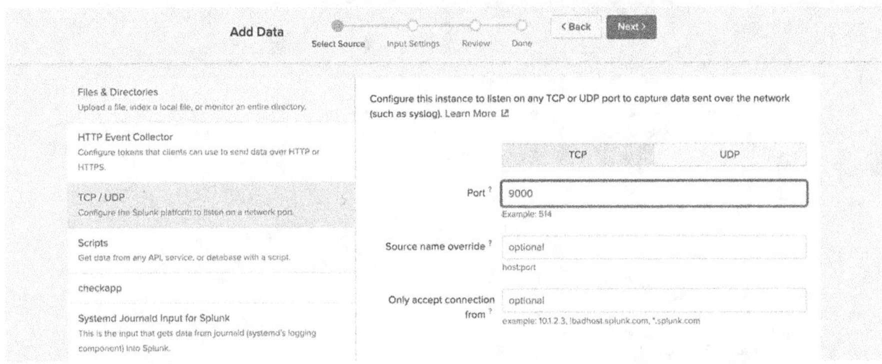

Figure 10-5. *TCP network input listening port 9000*

3. For advanced input settings, configure the form and
    provide the source type and index name. Refer to
    Figure 10-6 for help.

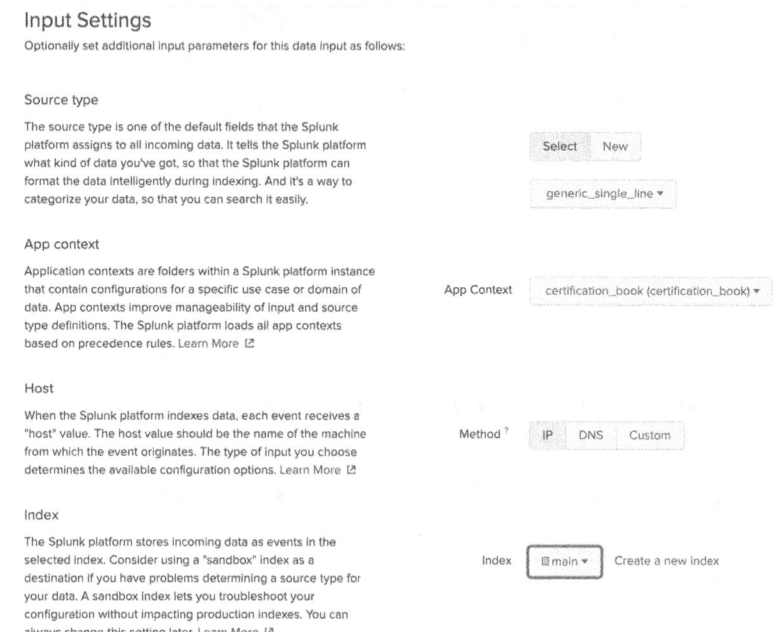

Figure 10-6. *Network input advanced input settings*

Modify Network Input Using .conf Files

You can modify network input using .conf files. You can change the index value, source value, and so forth using .conf files.

Configure TCP Network Input Using .conf File

To configure TCP network input, go to $SPLUNK_HOME/etc/apps/ certification_book/local/inputs.conf. In the inputs.conf stanza, [tcp://<host or ip>:<port>], there are various parameters that you can configure (see Table 10-4).

Table 10-4. *TCP network input inputs.conf*

Attribute	Value
host=<hostname>	Splunk uses the hostname during parsing and indexing.
index=<Index name>	Splunk sets an index to save an event.
sourcetype=<sourcetype name>	Source type in Splunk is used for formatting data during parsing and indexing.
source=<sourcename>	Source name in Splunk is an input where data originates.
queue = parsingQueue | indexQueue	Queue specifies where the input processor should deposit event that it reads.
connection_host=ip|dns	Ip in Splunk Enterprise sets the host to the IP address of the Splunk server. DNS in Splunk Enterprise sets the host to the DNS entry of the Splunk server.

If clients want to monitor all system logs listening on Transmission Control Protocol port 514, you can write a TCP stanza in inputs.conf, but Splunk should have root permissions to open a restricted port. Refer to the following example:

1. Go to $SPLUNK_HOME/etc/apps/certification_ book/local/inputs.conf. Create a stanza as follows:

```
[tcp://514]
connection_host=dns
sourcetype=syslog
index=main
```

2.   Save the changes.

3.   Restart the Splunk instance.

Configure Network UDP Input Using .conf File

To configure a UDP network input, go to $SPLUNK_HOME/etc/apps/ certification_book/local/inputs.conf. In the inputs.conf stanza, [udp://<host or ip>:<port>], there are various parameters that you can configure (see Table 10-5).

Table 10-5. UDP network input inputs.conf

attributeAttribute	Value
host=<hostname>	Splunk uses the hostname during parsing and indexing.
index=<Index name>	Splunk sets an index to save an event.
sourcetype=<sourcetype name>	Source type in Splunk is used for formatting data during parsing and indexing.
source=<sourcename>	Source name in Splunk is an input where data originates.
queue = parsingQueue l indexQueue	Queue specifies where the input processor should deposit event that it reads.
_rcvbuf=<value>	_rcvbuf in Splunk sets the receive buffer for UDP.
no_priority_stripping = true l false	no_priority_stripping sets how Splunk handles receiving Syslog data.
no_appending_timestamp = true l false	no_appending_timestamp sets how Splunk applies timestamps and hosts to events.

If clients want to monitor all Application Performance Monitoring (APM) logs listening on User Datagram Protocol port 9514, you can write a UDP stanza in inputs.conf. Refer to the following example:

1. Go to `$SPLUNK_HOME/etc/apps/certification_book/local/inputs.conf`. Create a stanza as follows:

```
[udp://9514]
connection_host=ip
sourcetype=apm
index=main
```

2. Save the changes.

3. Restart the Splunk instance.

Let's next discuss pulling data using agentless input.

Pulling Data Using Agentless Input

An HTTP Event Collector (HEC) is a token-based HTTP input that is secure and scalable. The HEC allows you to send data and application events to a Splunk Enterprise instance using HTTP and Secure HTTP (HTTPS) protocols. HEC uses a token-based authentication model.

An agent-based HTTP input is more secure and scalable and doesn't need to use forwarders.

HTTP Input Using Splunk Web

In this section, you create an HTTP event, configure the HEC, and modify a stanza using a configuration file. Then, you parse data from a local instance to a Splunk Enterprise instance using HEC tokens.

To get HTTP input using Splunk Web, refer to the following steps:

1. Go to Settings ➤ Data Inputs. Go to HTTP Event Collector and click New Token.

2. Enter the name **book_hec_input**.

3. Enter the source name and description (optional). If you want to enable indexer acknowledgment for this token, don't click the Enable indexer acknowledgment checkbox (in this exercise).

4. Click Next (see Figure 10-7).

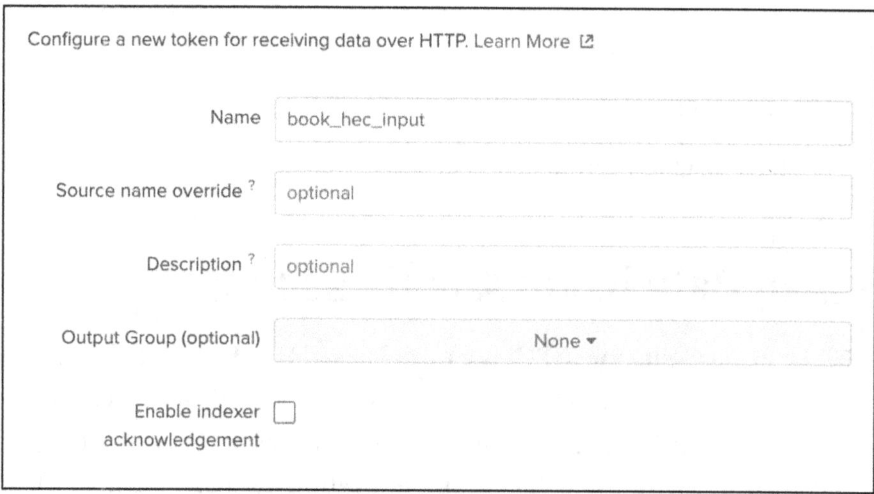

Figure 10-7. Configure book_hec_input HTTP input

5. (Optional) Select the source type and the transactions index for HEC events.

6. Click Review (see Figure 10-8).

Input Settings

Optionally set additional input parameters for this data input as follows:

Source type

The source type is one of the default fields that the Splunk
platform assigns to all incoming data. It tells the Splunk platform
what kind of data you've got, so that the Splunk platform can
format the data intelligently during indexing. And it's a way to
categorize your data, so that you can search it easily.

Automatic | Select | New

Index

The Splunk platform stores incoming data as events in the
selected index. Consider using a "sandbox" index as a
destination if you have problems determining a source type for
your data. A sandbox index lets you troubleshoot your
configuration without impacting production indexes. You can
always change this setting later. Learn More

Select Allowed Indexes | Available Item(s) — add all » | Selected item(s) « remove all

cim_modactions
history
main
summary
transactions

transactions

Select indexes that clients will be able to select from.

Default Index   transactions ▼   Create a new index

Figure 10-8. HTTP events advanced input settings

Configure HTTP Event Collector in Splunk

Before you can use an HTTP Event Collector in Splunk, you need to enable
it in global settings in Splunk Web. To configure the HEC to receive events
in Splunk Web, refer to the following instructions:

1. Click Settings and go to Data Inputs.

2. Click HTTP Event Collector.

3. Click Global Settings

4. In All tokens, toggle it to Enable.

5. Select Source Type, Index, and Output group
   (optional).

6. To use a deployment server to handle configurations
    for HEC tokens, click the Use Deployment Server
    check box.

7. To have HEC listen and communicate over HTTPS
    rather than HTTP, click the Enable SSL check box.
    (For this exercise, don't enable it.)

8. Enter the HTTP port number (whichever port
    number you want) to make it listen to HTTP events.

9. Click Save (see Figure 10-9).

Edit Global Settings   ✕

All Tokens	Enabled	Disabled
Default Source Type	Select Source Type ▾	
Default Index	Default ▾	
Default Output Group	None ▾	
Use Deployment Server	☐	
Enable SSL	☐	
HTTP Port Number ?	8088	

Cancel   Save

Figure 10-9. *HTTP Event Collector*

Enable HTTP Input Using .conf File

To enable HTTP input using a .conf file, go to $SPLUNK_HOME/etc/apps/
splunk_httpinput/local/inputs.conf. In inputs.conf, create a stanza
[http]. In that stanza, there are various parameters (see Table 10-6).

Table 10-6. *Configure HTTP input using inputs.conf*

Attribute	Value
disabled=<0|1>	When disabled=<0> HTTP input is enabled, and when disabled=<1> HTTP input is disabled.
host=<hostname>	Splunk uses the hostname during parsing and indexing.
index=<Index name>	Splunk sets an index to save an event.
sourcetype=<sourcetype name>	Source type in Splunk is used for formatting data during parsing and indexing.
source=<sourcename>	Source name in Splunk is an input where data originates.
enableSSL=0|1	When enableSSL=<0> SSL is off and when enable SSL =<1> SSL is enabled.

To update HTTP input using a direct configuration file in this exercise, enable SSL for HTTP input.

1. Go to $SPLUNK_HOME/etc/apps/splunk_httpinput/ local/inputs.conf. Create a stanza as follows:

```
[http]
disabled=0
enableSSL=1
```

2. Save the changes.

3. Restart the Splunk instance.

Configure HTTP Event Collector in Splunk Using .conf File

To configure an HTTP Event Collector to receive events in Splunk Web, you need to enable HEC (HTTP Event Collector) through global settings or using inputs.conf, as shown in the previous step.

1. Go to $SPLUNK_HOME/etc/apps/certification_ book/local/inputs.conf.

2. In inputs.conf, create a stanza, [http://<HEC Name>]. In that stanza, there are various parameters (see Table 10-7).

Table 10-7. *Configure HTTP Event Collector inputs.conf*

Attribute	Value
disabled=<0l1>	When disabled=<0> HTTP input is enabled, and when disabled=<1> HTTP input is disabled.
token=<value>	Token in HTTP helps in authentication, and the value of the token is Unique. Easy to identify. Token value is called a *globally unique identifier.*
host=<hostname>	Splunk Enterprise uses the hostname during parsing and indexing. A hostname is used at searching.
index=<Index name>	Splunk Enterprise sets an index where it should save an event.
Sourcetype=<sourcetype name>	The source type in Splunk Enterprise is used for formatting of data during parsing and indexing of data.
source=<sourcename>	The source name in Splunk Enterprise is an input file where data originates.
useACK=0l1	When useACK=<0> indexer acknowledgement is disabled, and when useAck =<1> indexer acknowledgement is enabled.

Next, let's modify the HTTP Event Collector in Splunk using the inputs.
conf file where you enable acknowledgement.

1. Go to `$SPLUNK_HOME/etc/apps/certification_`
   `book/local/inputs.conf`. Modify the stanza as
   follows:

   ```
   [http://json_api]
   disabled=0
   token=<define_token>
   useACK=1
   index=main
   ```

2. Save the changes.

3. Restart the Splunk instance.

Parse Data in Splunk Using HTTP Event Collector

Now, you can parse data to a Splunk instance using an HTTP Event
Collector. You can refer to the following code block and Figure 10-10 to
better understand how to parse data in a Splunk instance using HEC:

```
curl -k  https://localhost:8088/services/collector -H
"Authorization: Splunk <token>"  -d '{"sourcetype": "trial"
,  "event": "hello world"}'
```

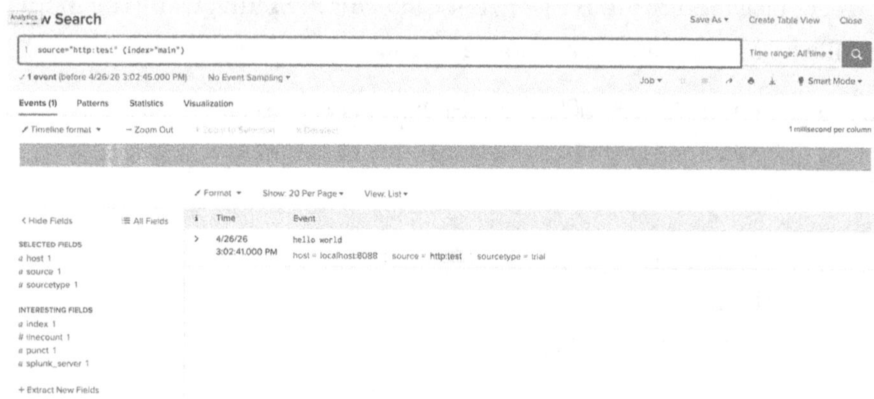

Figure 10-10. *HTTP Event Collector events*

You have reached the end of this chapter, so congratulations on learning about additional forwarding options, monitor input, scripted input, and network inputs, which are responsible for the bulk of data in the Splunk environment.

Summary

You learned how to get data from various sources, including monitor input, network input, scripted input, and agentless input. You can now enter data in a more defined manner while ensuring that the events are not intercepted or altered. You can also enable the security of the feed and control and modify the queue size. You now know that acknowledgment indexers are one of the features of the Splunk platform. You can check your understanding using the following questions.

Multiple-Choice Questions

A. Which of the following enables compression for universal forwarders in outputs.conf?

1. [udpout:mysplunk_indexer1] compression=true

2. [tcpout]defaultGroup=my_indexers compressed=true

3. /opt/splunkforwarder/bin/splunk enable compression

4. [tcpout:my_indexers] server=mysplunk_ indexer1:9997,mysplunk_indexer2:9997 decompression=false

B. A universal forwarder has which capabilities when sending data? (Select all that apply.)

1. Sending alerts

2. Compressing data

3. Obfuscating/hiding data

4. Indexer acknowledgment

C. Which of the following statements apply to directory inputs? (Select all that apply.)

1. All discovered text files can be consumed.

2. Compressed files are ignored by default.

3. Splunk recursively traverses through the directory structure.

4. When adding new log files to a monitored directory, the forwarder must be restarted to take them into account.

   D. What is the default size of the queue in Splunk with useACK=true?

      1. 500 KB.

      2. 7 MB.

      3. 1MB.

      4. There are no queues in Splunk with useACK=true.

   E. When configuring monitor input with whitelists or blacklists, what is the supported method of filtering the lists?

      1. Slash notation

      2. Regular expression

      3. Irregular expression

      4. Wildcard-only expression

   F. To set up a network input in Splunk, what needs to be specified?

      1. File path

      2. Username and password

      3. Network protocol and port number

      4. Network protocol and MAC address

   G. Which option accurately describes the purpose of the HTTP Event Collector (HEC)?

      1. A token-based HTTP input that is secure and scalable and that requires the use of forwarders

      2. A token-based HTTP input that is secure and scalable and that does not require the use of forwarders

3. An agent-based HTTP input that is secure and scalable and that does not require the use of forwarders

4. A token-based HTTP input that is insecure and non-scalable and that does not require the use of forwarders

H. What is the difference between the two wildcards ... and * for the monitor stanza in inputs.conf?

1. ... is not supported in monitor stanzas.

2. No difference—they are interchangeable and match anything beyond directory boundaries.

3. * matches anything in that specific directory path segment, whereas ... recurses through subdirectories as well.

4. ... matches anything in that specific directory path segment, whereas * recurses through subdirectories as well.

Answers

A. 2

B. 2, 4

C. 1, 3

D. 2

E. 2

F. 3

G. 2

H. 3

References

- https://docs.splunk.com/Splexicon:Scriptedinput

- https://help.splunk.com/en/splunk-enterprise/
  get-started/get-data-in/10.0/get-data-from-
  network-sources/get-data-from-tcp-and-udp-ports

- https://help.splunk.com/en/splunk-enterprise/
  developing-views-and-apps-for-splunk-
  web/10.0/build-scripted-inputs/setting-up-a-
  scripted-input

- https://help.splunk.com/en/splunk-enterprise/
  get-started/get-data-in/10.0/get-windows-data/
  monitor-windows-network-information

- https://help.splunk.com/en/splunk-enterprise/
  get-started/get-data-in/10.0/get-data-with-
  http-event-collector/set-up-and-use-http-event-
  collector-in-splunk-web

- https://help.splunk.com/en/splunk-enterprise/
  administer/manage-users-and-security/10.0/
  secure-splunk-platform-communications-
  with-transport-layer-security-certificates/
  configure-splunk-indexing-and-forwarding-to-
  use-tls-certificates

Splunk's Advanced .conf File and Diag

This chapter discusses the various .conf files responsible for storing Splunk configuration information and apps. You learn about fine-tuning the input. You study the process of anonymizing sensitive personal data and learn about debugging configuration files. You are introduced to the diag, which collects basic information on the environment, the instance, and the operating system.

In a nutshell, the following are topics covered in this chapter:

- Understanding Splunk .conf files
- Setting fine-tuning input (custom source type)
- Anonymizing data
- Understanding merging logic in Splunk
- Debugging a configuration file
- Creating a diag

© Carlos Moreno Buitrago, Deep Mehta 2026
C. M. Buitrago and D. Mehta, *The Splunk Core User Study Companion*, Certification Study Companion Series, https://doi.org/10.1007/979-8-8688-2501-9_11

Understanding Splunk .conf files

As you noticed in previous chapters, the configuration file is called .conf file in Splunk. A Splunk configuration file contains Splunk Enterprise configuration information.

Default configuration files are saved in $SPLUNK_HOME/etc/system/default. (It is advised to edit local files in Splunk and avoid editing default files because they are overridden whenever Splunk is updated.) To edit local files, go to $SPLUNK_HOME/etc/system/local. Stanzas in the configuration files include system settings, authentication and authorization information, index-related settings, deployment, cluster configuration, knowledge objects, and searches in Splunk.

The following files are discussed in this section:

- props.conf

- indexes.conf

- transforms.conf

- inputs.conf

- outputs.conf

- deploymentclient.conf

props.conf

The props.conf file is present on the forwarders, search head, and indexer. In this file, you can apply parsing rules on data. The props.conf file is mainly used for the following:

- Configuring line breaking for events

- Character encoding

- Processing of binary files

- Timestamp recognition and converting time formats

- Event segmentation

- Field extractions and calculations

The props.conf file is located in `$SPLUNK_HOME/etc/system/local/props.conf`.

Table 11-1 describes the file's attributes and their respective values.

Table 11-1. *Configuring props.conf*

Attribute	Value
SHOULD_LINE_MERGE=false|true	It specifies whether to combine several lines of data into a single multiline event.
TIME_PREFIX=<regular expression>	The regex of an event checks for the timestamp in an event.
TIME_FORMAT=<strptime-style format>	It specifies the "strptime" format string and extracts the date from the event.
MAX_TIMESTAMP_LOOKAHEAD=<Integer>	Specifies the number of characters that Splunk should check in a timestamp.
LINE_BREAKER=<regular expression>	It identifies the start of the next event.
BREAK_ONLY_BEFORE=<regular expression>	It is a method for defining the start of the next event.
MAX_EVENTS=<Integer>	Specifies the maximum number of lines for an event.
ANNOTATE_PUNCT=false|true	Enables or disables punctuation in an event for searching.
KV_MODE = <none|KV pairs>	It is used to define the KV pairs.
SEDCMD-<class> = s/<regex>/<replacement>/g	It hides data and can be replaced with other data.

Table 11-1 consists of a few of the important instructions that are used in props.conf. For more information and options, refer to `https://help.splunk.com/en/splunk-enterprise/administer/admin-manual/10.0/configuration-file-reference/10.0.1-configuration-file-reference/props.conf`.

indexes.conf

The indexes.conf file exists on the Splunk indexer mainly to configure indexes and manage index policies, such as data expiration and data thresholds. The file is primarily used for configuring indexes and their properties. The indexes.conf file is located in `$SPLUNK_HOME/etc/system/local/indexes.conf` or in a specific app to manage the indexes.

Table 11-2 describes the options to use in the indexes.conf file.

Table 11-2. *Configuring indexes.conf*

Attribute	Value
homePath=<Path>	homePath contains the location of hot and warm buckets.
coldPath=<Path>	coldPath contains the location of cold buckets.
thawedPath=<Path>	thawedPath contains the location of thawed database.
repFactor=<int >	Determines whether the indexer needs to be replicated or not.
maxHotBuckets=<int>	The maximum number of hot buckets that can exist in the index.
maxDataSize=<int>	It specifies the maximum size of hot buckets.
maxWarmDBCount=<int>	It specifies the maximum number of warm buckets.
maxTotalDataSizeMB=<int>	It specifies the maximum size of a particular index.
frozenTimePeriodInSecs=<int>	The number of seconds after which the index rolls to frozen.
coldToFrozenDir=<Path>	It specifies a path for the frozen archive.

Table 11-2 lists some of the important instructions in indexes.conf. For more information, refer to `https://help.splunk.com/en/splunk-enterprise/administer/admin-manual/10.0/configuration-file-reference/10.0.1-configuration-file-reference/indexes.conf`.

transforms.conf

The transforms.conf file is present on the heavy forwarder, search head, and indexer. In this file, you can apply the rules for parsing data and applying regular expression rules to transform source type, anonymize certain types of sensitive incoming data, and create advanced search field extraction. The transforms.conf file is located in $SPLUNK_HOME/etc/system/local/transforms.conf or, as a recommendation, in a specific app to manage the rules for each data source.

Table 11-3 describes how to configure the transforms.conf file.

Table 11-3. *Configuring transforms.conf*

Attribute	Value
SOURCE_KEY=<_raw>	It indicates which data stream to use as the source for pattern matching.
REGEX=<regular expression>	REGEX identifies events from SOURCE_KEY.
DEST_KEY=<Metadata:Sourcetype>	DEST_KEY indicates where to write the processed data.
FORMAT=<Sourcetype::<sourcetype name>>	FORMAT controls how REGEX writes the DEST_KEY.

Table 11-3 lists a few of the important instructions in transforms.conf. For more information, go to https://help.splunk.com/en/splunk-enterprise/administer/admin-manual/10.0/configuration-file-reference/10.0.1-configuration-file-reference/transforms.conf.

inputs.conf

The inputs.conf file exists on the universal forwarder, heavy forwarder, search head, and indexer. However, there are various attributes in the inputs.conf file that need to be handled before inputting data. The inputs.conf file is located in $SPLUNK_HOME/etc/system/local/inputs.conf or, as a recommendation, in a specific app to manage the inputs for each data source type.

Table 11-4 illustrates the configuration of inputs.conf file.

Table 11-4. *Configuring inputs.conf*

Attribute	Value
host=<string>	It sets the host value to the static value for the input stanza.
index=<string>	It specifies the index to store events that come through the input stanza.
source=<string>	It sets the source field for the events from the input field.
sourcetype=<string>	It sets the source type field for the events from these input fields.
host_regex=<regular expression>	The regular expression extracts the host from the path to the host file.
whitelist=<regular expression>	The files from this input are monitored if the path matches the regular expression.
blacklist=<regular expression>	The files from this input are not monitored if the path matches the regular expression.
outputgroup=<string>	The name of the group where the event collector forwards data.
enableSSL=0|1	It decides whether the HTTP event collector group should use SSL or not.
maxSockets=<integer>	It decides the count of the HTTP connections that the HTTP event collector input accepts simultaneously.
queue=parsingQueue|indexQueue	It tells the index where it should submit the incoming events for a particular input stanza.

Table 11-4 lists a few of the important instructions. For more information on inputs.conf, refer to `https://help.splunk.com/en/ splunk-enterprise/administer/admin-manual/10.0/configuration- file-reference/10.0.1-configuration-file-reference/inputs.conf`.

outputs.conf

The outputs.conf file is present on the universal forwarder, the heavy forwarder, and the search head. In this file, you can apply rules for sending data out to Splunk instances. However, various attributes need to be handled before the forwarder sends data to the receiving Splunk instances. The outputs.conf file is located in `$SPLUNK_HOME/etc/system/local/ outputs.conf`.

Table 11-5 illustrates the configuration of the outputs.conf file.

Table 11-5. *Configuring outputs.conf*

Attribute	Value
heartbeatFrequency=<integer>	Heartbeat frequency notifies the receiving server by sending a heartbeat package specified in the integer block.
maxQueueSize=<integer>	It indicates the RAM size of all items in the queue.
autoLBFrequency=<integer>	It is used for load balancing, specifying the time in the integer block to change the server.
autoLBVolume=<integer>	The volume of data in bytes to send to an indexer before the new one is selected from available indexers.

(continued)

Table 11-5. (*continued*)

Attribute	Value
maxEventSize=<integer>	The maximum size of an event that Splunk can transmit.
server=<ip>:<port>,<servername>:<port>	It specifies the target indexer for transmitting the data.
timestampformat=<format>	It specifies the prepended formatted timestamp format for events.
manager_uri=<ip>:<port>,<servername>:<port>	The address of the manager of the cluster.
type=tcpludp	It specifies whether you want to use a TCP or UDP protocol.
compressed=truelfalse	It specifies whether data needs to be compressed or not.
_TCP_ROUTING=<target_group>	This specifies the target group to forward data.
disabled=truelfalse	It determines whether it needs to disable events transmitting or not.

Table 11-5 lists a few of the important instructions in outputs.conf. For more information, refer to `https://help.splunk.com/en/splunk-enterprise/administer/admin-manual/10.0/configuration-file-reference/10.0.1-configuration-file-reference/outputs.conf`.

deploymentclient.conf

The deploymentclient.conf file resides on the universal forwarder and the heavy forwarder if they are managed by the deployment server. This file contains descriptions of the settings that you can use to customize a deployment client's behavior. There are various attributes in the deploymentclient.conf files that are useful for setting the client–server communication. The deploymentclient.conf file is located in `$SPLUNK_HOME/etc/system/local/deploymentclient.conf` or, as a recommendation, in a specific app to manage the forwarder configuration.

Table 11-6 shows the various attributes and their values in the deploymentclient.conf file.

Table 11-6. *Configuring deploymentclient.conf*

Attribute	Value
targetUri=<uri>	Specifies the address of the deployment server.
connect_timeout=<integer>	The maximum time in seconds that a deployment client can take to connect to the deployment server.
send_timeout=<integer>	The maximum time in seconds that a deployment client can take to send or write data to the deployment server.
recv_timeout=<integer>	The maximum time in seconds that a deployment client can take to read or receive data to the deployment server.

Table 11-6 lists a few of the important instructions in deploymentclient.conf. For more information, refer to `https://help.splunk.com/en/splunk-enterprise/administer/admin-manual/10.0/configuration-file-reference/10.0.1-configuration-file-reference/deploymentclient.conf`.

You have completed the discussion on the six configuration files in Splunk. In the next section, you learn about the fine-tuning inputs that create customized source types for input data.

Setting Fine-Tuning Input

The fine-tuning inputs in Splunk are responsible for creating custom source types for data, which are scripted inputs, unusual log files, custom data types, and others. You can also create your source types by modifying props.conf or in Splunk Web so that Splunk can understand all that data. At the end of the day, Splunk is an observability tool which can ingest all kind of data. The fine-tuning process will help you to better understand the data and get ready the field extractions you need for your searches, reports, and dashboards.

Custom Source Types Using Splunk Web

In this section, you create a custom source type for the client_data.log file, used in Chapter 1, so that Splunk can recognize events, extract the timestamp from events, and understand where a new event is starting. To do this for your data using Splunk Web, refer to the following instructions:

1. Go to Settings and head over to Source types.

2. Search by "client_data", or create a new source type. The sourcetype will be defined in the certification_ book app.

3. In Category, select Custom Data.

4. Splunk has predefined indexed extraction based on the type of data you can select. For test.txt, select none. (For a better understanding, refer to Figure 11-1.)

Edit Source Type: client_data   ×

Description	optional
Destination app	certification_book ▾
Category	Custom ▾
Indexed extractions ?	none ▾

Figure 11-1. *Custom sourcetype client_data*

5. In Event Breaks, there are three types of event-
    breaking policies: Auto, Every Line, and Regex.
    Select according to your data, but for this example,
    we will use:

    ```
    ([\r\n]+)\d+\s+\"\$EIT\,
    ```

6. In the Timestamp tab, there are four policies to
    determine timestamps for incoming data: Auto,
    Current Time, Advanced, and Configuration.

7. Under Advanced, add TIME_FORMAT, MAX_
    TIMESTAMP, TRUNCATE:

```
SHOULD_LINEMERGE=false
LINE_BREAKER= ([\r\n]+)\d+\s+\"\$EIT\,
NO_BINARY_CHECK=true
category=Custom
TIME_FORMAT = %m/%d/%Y %k:%M
TIME_PREFIX = -\s+\d{5}\s+
MAX_TIMESTAMP_LOOKAHEAD = 16
TRUNCATE = 5000
```

Custom Source Types Using props.conf

Creating a customized source type is the easiest and best option when using a props.conf file. Go to $SPLUNK_HOME/etc/apps/certification_book/local/props.conf. There are six crucial configurations to define any custom source type in Splunk. Many policies are added and subtracted, totally depending on the input data, but overall, these six policies are important.

The configurations are defined in the following code block:

```
SHOULD_LINEMERGE=false|true
TIME_PREFIX=<regular expression>
TIME_FORMAT=<strptime-style format>
MAX_TIMESTAMP_LOOKAHEAD=<Integer>
LINE_BREAKER=<regular expression>
TRUNCATE=<Integer>
```

To create a customized source type for client_data.log, where the source type is named [client_data], follow these steps in props.conf:

```
[client_data]
SHOULD_LINEMERGE=false
LINE_BREAKER= ([\r\n]+)\d+\s+\"\$EIT\,
NO_BINARY_CHECK=true
category=Custom
TIME_FORMAT = %m/%d/%Y %k:%M
TIME_PREFIX = -\s+\d{5}\s+
MAX_TIMESTAMP_LOOKAHEAD = 16
TRUNCATE = 5000
```

You have completed the creation of source types using fine-tuning input and the use of props.conf.

Note   If you create source types in Splunk Cloud using Splunk Web, Splunk Cloud manages the source type configurations automatically. However, if you have Splunk Enterprise and manage a distributed configuration, you must distribute a new source type.

The next section discusses the process of anonymizing sensitive data and credentials.

Anonymizing the Data

It is essential to mask sensitive personal information such as credit card numbers and social security numbers. You can anonymize parts of data in events to protect privacy while providing the remaining data for analysis.

There are two ways to anonymize data in Splunk:

- Use props.conf to anonymize data with a sed script.

- Use props.conf and transforms.conf to anonymize data with regular expressions.

props.conf to Anonymize Data with a sed Script

In this process, you directly use props.conf to anonymize parts of data in events to protect privacy while providing the remaining data for analysis. The data can be anonymized using a sed (stream editor) script to replace a substitute string in an event. Sed is a *nix utility that reads a file and modifies the input based on commands within or arguments supplied to the utility. You can use sed like syntax in props.conf to anonymize data.

Syntax to Anonymize Data with a sed Script

```
SEDCMD-<class> = s/<string1>/<string2>/
```

In client_data.log, there is data that looks similar to the data in the following code block:

```
1 "$EIT,907409,38550,E,0,,0,1,0,0,16777317,0,8,A,19.079747,72.
849640,65529,195,183023,261218,23,1,15,0,4208,1148,0,1,0,0,0,4.
2,E10.21,0" 27.97.83.90 - 58840 12/27/2018 0:00
```

You need to mask the IP address to XXXX. You need to first modify inputs.conf, which has a few prerequisite policies similar to the policy in the following code block:

inputs.conf
```
[monitor://desktop/client_data.log]
sourcetype=client_data
```

After updating inputs.conf, you can update `props.conf`. In props.conf, write SEDCMD to mask data. The SEDCMD command to mask IPs is similar to the following policy:

```
SEDCMD-hideip=s/(\d{1,3}\.\d{1,3}\.\d{1,3}\.\d{1,3})/ ip=xxxx/g
```

After the SEDCMD command, the props.conf for the client_data source type is similar to the following block:

```
[client_data]
SHOULD_LINEMERGE=false
LINE_BREAKER= ([\r\n]+)\d+\s+\"\$EIT\,
NO_BINARY_CHECK=true
category=Custom
TIME_FORMAT = %m/%d/%Y %k:%M
TIME_PREFIX = -\s+\d{5}\s+
```

```
MAX_TIMESTAMP_LOOKAHEAD = 16
TRUNCATE = 5000
SEDCMD- SEDCMD-hideip=s/(\d{1,3}\.\d{1,3}\.\d{1,3}\.\d{1,3})/
ip=xxxx/g
```

Once the props.conf is updated for the client_data source type, the IP address is masked for new events. The sedcmd command anonymizes the data at index time. Example:

```
"$EIT,907409,38550,E,0,,0,1,0,0,16777317,0,8,A,19.079747,72.84
9640,65529,195,183023,261218,23,1,15,0,4208,1148,0,1,0,0,0,4.2
,E10.21,0" ip=xxxx - 58840 12/27/2018 0:00
```

Let's look at how to anonymize data using props.conf and transforms. conf with regular expressions.

props.conf and transforms.conf to Anonymize Data with Regular Expressions

Let's use props.conf and transforms.conf to anonymize parts of data in events to protect privacy while providing remaining data. transforms. conf and props.conf transform events if the input event matches a regular expression. Let's use the previous example with client_data, but calling transforms.conf for masking data. The props.conf block looks similar to the following block:

```
[client_data]
SHOULD_LINEMERGE=false
LINE_BREAKER= ([\r\n]+)\d+\s+\"\$EIT\,
NO_BINARY_CHECK=true
category=Custom
TIME_FORMAT = %m/%d/%Y %k:%M
TIME_PREFIX = -\s+\d{5}\s+
```

```
MAX_TIMESTAMP_LOOKAHEAD = 16
TRUNCATE = 5000
SEDCMD- SEDCMD-hideip=s/(\d{1,3}\.\d{1,3}\.\d{1,3}\.\d{1,3})/
ip=xxxx/g
TRANSFORMS-anonymize = ip-anonymizer
```

After updating props.conf, you need to update transforms.conf so that its block is similar to the following block:

```
[ip-anonymizer]
REGEX = (.*)(\d{1,3}\.\d{1,3}\.\d{1,3}\.\d{1,3})(.*)
FORMAT = $1ip=xxxx$3
DEST_KEY = _raw
```

Now, let's turn to merging logic in Splunk.

Understanding Merging Logic in Splunk

In Splunk, there are various configuration files located in different places. With merging logic, all these configuration files are brought together and combined in one global file. In this section, you learn how merging is performed and get to know how Splunk combines configuration files.

Configuration File Precedence

Splunk uses configuration files to determine every aspect of its behavior. Configuration file precedence merges the settings from all copies of the file. Different copies can have a conflicting attribute value, so the file with the highest priority determines the configuration file's priority by location in the directory structure.

Splunk Determine Precedence Order

Splunk determines the precedence order based on the following rules:

- System local directory—highest priority

- App local directories

- App default directories

- System default directory—lowest priority

Splunk .conf Files Location

Splunk configuration files reside in $SPLUNK_HOME/etc.

The following directories contain Splunk configuration files:

- **$SPLUNK_HOME/etc/system/default**: This directory contains the default configuration files in Splunk. The default directory contains preconfigured versions of the configuration files. Never edit default files because they will be overwritten during an upgrade.

- **$SPLUNK_HOME/etc/system/local**: This directory contains the local configuration files. You can create and edit your files in one of these directories. These directories are not overwritten during updates.

- **$SPLUNK_HOME/etc/apps/<app_name>/local**: This directory contains the local configuration for applications in Splunk. All configurations for your application should be placed here.

- **$SPLUNK_HOME/etc/users**: This is the user-specific configuration directory. If you want to change any user-specific configuration, you update it here.

- **$SPLUNK_HOME/etc/system/README**: This directory contains the reference files. Reference files are either .example files or .spec files.

Configuration Merging Logic

Configuration merging logic is simple. To merge, the configurations' names, stanza names, and attribute names should match.

Merging in Splunk is done based on configuration file precedence.

Example 1: Configuration Merging (No Conflict)

Let's look at an example of configuration merging logic with no conflicts.

You have $SPLUNK_HOME/etc/system/local/props.conf.

```
[client_data]
MAX_TIMESTAMP_LOOKAHEAD = 16
TRUNCATE = 5000
```

And you have $SPLUNK_HOME/etc/apps/certification_book/local/props.conf.

```
[client_data]
SHOULD_LINEMERGE=false
```

The resulting configuration has no conflicts, so Splunk merges the files.

```
[client_data]
MAX_TIMESTAMP_LOOKAHEAD = 16
TRUNCATE = 5000
SHOULD_LINEMERGE=false
```

Example 2: Configuration Merging (Conflict)

The following is for $SPLUNK_HOME/etc/system/local/props.conf.

```
[client_data]
MAX_TIMESTAMP_LOOKAHEAD = 16
TRUNCATE = 5000
```

You now merge with $SPLUNK_HOME/etc/apps/certification_book/local/props.conf, which has the following configuration:

```
[client_data]
TRUNCATE = 1000
```

The resulting configuration, $SPLUNK_HOME/etc/apps/certification_book/local/props.conf, is unable to overwrite $SPLUNK_HOME/etc/system/local/props.conf because in file precedence, files in the system local folder have greater priority than files in apps. The resulting configuration is as follows:

```
[client_data]
MAX_TIMESTAMP_LOOKAHEAD = 16
TRUNCATE = 5000
```

Example 3: Configuration Merging (Conflict)

You have $SPLUNK_HOME/etc/apps/search/default/props.conf.

```
[client_data]
MAX_TIMESTAMP_LOOKAHEAD = 16
TRUNCATE = 5000
```

And you want to merge $SPLUNK_HOME/etc/apps/search/local/props.conf.

```
[client_data]
TRUNCATE = 9999
```

The resulting configuration, $SPLUNK_HOME/etc/apps/search/local/ props.conf, can overwrite the value of $SPLUNK_HOME/etc/apps/search/ default/props.conf because in file precedence, the files in local have greater priority than files in default. Hence, the resulting configuration is as follows:

```
[client_data]
MAX_TIMESTAMP_LOOKAHEAD = 16
TRUNCATE = 9999
```

Let's now discuss debugging a configuration file using btool.

Debugging Configuration Files

Btool in Splunk is used for displaying merged on-disk configurations. Btool is a command-line tool that can help troubleshoot configuration file issues or see what values are being used by your Splunk Enterprise installation. Splunk provides btool in the $SPLUNK_HOME/bin directory. Btool displays the on-disk configuration file settings, and if you change a setting, then you do not need to restart Splunk.

The syntax for btool <conf_file_prefix> is as follows:

```
./splunk cmd btool <conf_file_prefix> list
```

In the preceding syntax, <conf_file_prefix> stands for the name of the configuration you are interested in.

The syntax for btool <conf_file_prefix> is as follows:

```
./splunk cmd btool server list --debug
```

The preceding syntax determines which server.conf stanzas are being recognized.

Example: btool for Troubleshooting a Configuration File

This is the configuration of $SPLUNK_HOME/etc/apps/certification_
book/local/inputs.conf.

```
[monitor://desktop/client_data.log]
sourcetype=client_data
```

Change the configuration to the following code:

```
[monitor://desktop/client_data.log]
ourcetype=client_data
```

To troubleshoot the configuration file, go to $SPLUNK_HOME/bin.

```
./splunk cmd btool check
```

Note that btool prompts you with an error: invalid key in the stanza. A similar error is shown in Figure 11-2.

```
sh-3.2# splunk btool check
            Invalid key in stanza [monitor://desktop/client_data.log] in /Applications/Splunk
/etc/apps/certification_book/local/inputs.conf, line 2: ourcetype (value: client_data).
```

Figure 11-2. *btool error check for client_data*

Let's now look at creating a diag.

Creating a Diag

In Splunk, a diag is used for collecting basic information regarding your Splunk environment, instance, configuration file, and so forth. It gathers information regarding server specification, OS version, file system, and current open connections from the node running in the Splunk environment. A diag can contain app configurations, internal Splunk log files, and index metadata from the Splunk instance. Diags are stored in `$SPLUNK_HOME/var/run/diags`.

Creating a Diag in Splunk

Diags can be created using Splunk Web or a terminal/CLI. Starting in Splunk 6.0, you can generate diag for remote instances, but you need to have at least one of the following server roles:

- A search head, which is the only search head in a deployment

- A clustered search head

- A clustered indexer

- An indexer cluster manager

To create a diag using the command line, go to `$SPLUNK_HOME/bin` and run the following code:

```
./splunk diag
```

Figure 11-3 shows the diag created for a local instance.

```
(base) Deeps-MacBook-Air:bin deepmehta$ sudo ./splunk diag
Collecting components: conf_replication_summary, consensus, dispatch, etc, file_validate, index_files, index_listing, kvstore, log, searchpeers, suppression_listing
Skipping components: rest
Selected diag name of: diag-Deeps-MacBook-Air.local-2019-11-16_18-39-31
Starting splunk diag...
Logged search filtering is enabled.
Skipping REST endpoint gathering...
Determining diag-launching user...
Getting version info...
Getting system version info...
Getting file integrity info...
Getting network interface config info...
Getting splunk processes info...
Getting netstat output...
Getting info about memory, ulimits, cpu (on windows this takes a while)...
Getting etc/auth filenames...
Getting Sinkhole filenames...
Getting search peer bundles listings...
Getting conf replication summary listings...
Getting suppression files listings...
Getting KV Store listings...
Getting index listings...
Copying Splunk configuration files...
filtered out file '/Applications/Splunk/etc/apps/splunk_archiver/java-bin/jars/vendors/spark/2.3.3/lib/spark-core_2.11-2.3.3.jar'  limit: 10485760  size: 13123587
filtered out file '/Applications/Splunk/etc/apps/splunk_archiver/java-bin/jars/thirdparty/hive_1_2/hive-exec-1.2.1.jar'  limit: 10485760  size: 20599029
filtered out file '/Applications/Splunk/etc/apps/splunk_archiver/java-bin/jars/thirdparty/aws/aws-java-sdk-1.10.8.jar'  limit: 10485760  size: 21006573
The following certificates were excluded from the diag output automatically.
        /Applications/Splunk/etc/auth/appsCA.pem
        /Applications/Splunk/etc/auth/server.pem
        /Applications/Splunk/etc/auth/cloudCA.pem
        /Applications/Splunk/etc/auth/appsLicenseCA.pem
        /Applications/Splunk/etc/auth/ca.pem
        /Applications/Splunk/etc/auth/cacert.pem
        /Applications/Splunk/etc/auth/distServerKeys/trusted.pem
        /Applications/Splunk/etc/auth/distServerKeys/private.pem
        /Applications/Splunk/etc/auth/audit/public.pem
        /Applications/Splunk/etc/auth/audit/private.pem
        /Applications/Splunk/etc/auth/splunkweb/cert.pem
        /Applications/Splunk/etc/auth/splunkweb/privkey.pem
If you have any certs that were not auto-detected, please add them to an EXCLUDE rule in the [diag] stanza of server.conf.
Copying Splunk log files...
Copying bucket info files...
Copying Splunk dispatch files...
Copying Splunk consensus files...
Adding manifest files...
Adding cachemanager_upload.json...
Cleaning up...
Splunk diagnosis file created: /Applications/Splunk/diag-Deeps-MacBook-Air.local-2019-11-16_18-39-31.tar.gz
```

Figure 11-3. *Splunk diag*

To exclude the file from diag, run the following code:

```
./splunk diag --exclude "*/passwd"
```

When a Splunk diag is run, it produces a tar.gz file and a diag.log file.

You have now reached the end of the chapter. You can check your knowledge using the questions provided after the Summary.

Summary

This chapter discussed the various configuration files existing in Splunk: inputs.conf, outputs.conf, props.conf, transforms.conf, deploymentclient.conf, and indexes.conf. You learned about their key/value attributes and their utility in creating and modifying various commands. These files are responsible for storing enterprise and app configuration information. You also learned about custom source type creation using both Splunk Web and props.conf. You learned the process of anonymizing the sensitive or private data and the process of merging the configuration files located in several places into a single global file. In the end, you learned about debugging configuration files and the process of creating a diag that collects basic data like the environment instance, server specification, and open connections from the nodes. The chapter focused on configuration files, which are a crucial element of Splunk Enterprise.

According to the Splunk Enterprise Certified Admin exam blueprint, you covered the following modules in this chapter:

- Module 3: Splunk Configuration Files (5%)

- Module 7: Get Data In (5%)

- Module 15: Fine-Tuning Inputs (5%)

- Module 17: Manipulating Raw Data (5%)

The next chapter is a Splunk admin mock exam to help you get a better idea of the Splunk admin exam patterns.

Multiple-Choice Questions

A.   Default configuration files in Splunk are saved in which location?

   1. $SPLUNK_HOME/etc/system/local

   2. $SPLUNK_HOME/etc/bin/default

   3. $SPLUNK_HOME/etc/bin/local

   4. $SPLUNK_HOME/etc/system/default

B.   The props.conf file applies rules to read files.

   1. True

   2. False

C.   Where does the inputs.conf file reside? (Select all that apply.)

1. Universal forwarder

2. Heavy forwarder

3. Search head

4. Indexer

5. None of the above

D.   If you set the source type in inputs.conf for a given source, you cannot override the source type value in props.conf and transforms.conf.

1. True

2. False

E.   In which Splunk configuration is the SEDCMD command used?

1. props.conf

2. inputs.conf

3. transforms.conf

4. outputs.conf

F.   According to Splunk merging logic, which file has the highest priority?

1. System default directory

2. App local directory

3. App default directory

4. System local directory

Answers

A. 4

B. 2

C. 1, 2, 3, 4

D. 2

E. 1

F. 4

References

https://docs.splunk.com/Documentation/
Splunk/latest/Data/Anonymizedata.

https://docs.splunk.com/Documentation/
Splunk/latest/Admin/Propsconf

https://docs.splunk.com/Documentation/
Splunk/latest/Admin/Transformsconf

Splunk Admin Exam Set

This chapter presents multiple-choice questions useful for the Splunk Enterprise Certified Admin exam to give you an idea of what appears on the exams.

Questions

A. Default configuration files in Splunk are saved in which location?

1. $SPLUNK_HOME/etc/system/local

2. $SPLUNK_HOME/etc/bin/default

3. $SPLUNK_HOME/etc/bin/local

4. $SPLUNK_HOME/etc/system/default

B. Which setting in indexes.conf allows data retention to be controlled by time?

1. maxDaysToKeep

2. moveToFrozenAfter

3. maxDataRetentionTime

4. frozenTimePeriodInSecs

© Carlos Moreno Buitrago, Deep Mehta 2026
C. M. Buitrago and D. Mehta, *The Splunk Core User Study Companion*, Certification Study Companion Series, https://doi.org/10.1007/979-8-8688-2501-9_12

C. For a single line event source, it is most efficient to set SHOULD_LINEMERGE to what value?

1. true

2. false

3. <regex string>

4. newline character

D. The universal forwarder has which capabilities when sending data? (Select all that apply.)

1. Sending alerts

2. Compressing data

3. Obfuscating/hiding data

4. Indexer acknowledgement

E. When configuring monitor inputs with whitelists or blacklists, what is the supported method for filtering lists?

1. Slash notation

2. Regular expressions

3. Irregular expressions

4. Wildcards only

F. Which hardware attribute needs to be changed to increase the number of simultaneous searches (ad hoc and scheduled) on a single search head?

1. Disk

2. CPUs

3. Memory

4. Network interface cards

G. Which valid bucket types are searchable? (Select all that apply.)

   1. Hot buckets

   2. Cold buckets

   3. Warm buckets

   4. Frozen buckets

H. Which of the following are supported options when configuring optional network inputs?

   1. Metadata override, sender filtering options, network input queues (quantum queues)

   2. Metadata override, sender filtering options, network input queues (memory/persistent queues)

   3. Filename override, sender filtering options, network output queues (memory/persistent queues)

   4. Metadata override, receiver filtering options, network input queues (memory/persistent queues)

I. Which Splunk forwarder type allows parsing of data before forwarding to an indexer?

   1. Universal forwarder

   2. Parsing forwarder

   3. Heavy forwarder

   4. Advanced forwarder

J. What is the default character encoding used by Splunk during the input phase?

1. UTF-8

2. UTF-16

3. EBCDIC

4. ISO 8859

K. In which Splunk configuration is the SEDCMD used?

1. props.conf

2. inputs.conf

3. indexes.conf

4. transforms.conf

L. In which scenario would a Splunk administrator want to enable data integrity check when creating an index?

1. To ensure that hot buckets are still open for writes and have not been forced to roll to a cold state

2. To ensure that configuration files have not been tampered with for auditing and/or legal purposes

3. To ensure that user passwords have not been tampered with for auditing and/or legal purposes

4. To ensure that data has not been tampered with for auditing and/or legal purposes

M. Which Splunk component distributes apps and certain other configuration updates to search head cluster members?

1. Deployer

2. Cluster manager

3. Deployment server

4. Search head cluster manager

N. You need to monitor every file named error.log anywhere under /opt/apps, regardless of folder depth. Which stanza path is correct?

1. [monitor:///opt/apps/.../error.log]

2. [monitor:///opt/apps/*/error.log]

3. [monitor:///opt/apps/error.log]

4. [monitor:///opt/apps/*/*/error.log]

O. Which of the following are methods for adding inputs in Splunk? (Select all that apply.)

1. CLI

2. Splunk Web

3. editing inputs.conf

4. editing monitor.conf

P. Local user accounts created in Splunk store passwords in which file?

1. $SPLUNK_HOME/etc/users/passwd.conf

2. $SPLUNK_HOME/etc/authentication

3. $SPLUNK_HOME/etc/passwd

4. $SPLUNK_HOME/etc/users/authentication.conf

Q. What are the minimum required settings when creating a network input in Splunk?

1. Protocol, port number

2. Protocol, port, location

3. Protocol, username, port

4. Protocol, IP, port number

R. Which Splunk indexer operating system platform is supported when sending logs from a Windows universal forwarder?

1. Windows is not compatible with the UF.

2. Linux platform only.

3. Windows platform only.

4. Any OS platform.

S. You want to confirm the final, merged settings for the apache stanza in props.conf and see which file supplied them. Which command is best?

1. splunk btool props list --app apache

2. splunk btool props list apache --debug

3. splunk cmd btool --file props apache

4. splunk show config props apache --debug

Answers

A. 4

B. 4

C. 2

D. 2, 4

E. 2

F. 2

G. 1, 2, 3

H. 2

I. 3

J. 1

K. 1

L. 4

M. 1

N. 1

O. 1, 2, 3

P. 3

Q. 1

R. 4

S. 2

Summary

This chapter focused on multiple-choice questions useful for admin certification. You have come to the end of Module 2, which covered all the chapters that address the Splunk Enterprise Certified Admin certification. The next chapter begins the discussion on advanced Splunk deployment.

PART III

Advanced Splunk

Infrastructure Planning with Indexer and Search Head Clustering

This module deals with the advanced Splunk deployment. Deployment acts as a centralized configuration manager for instances; network-based installations do not require administrators to individually install each operating system.

The first chapter in this module discusses the architecture of Splunk. It also covers clustering. The following topics are discussed in this chapter:

- Capacity planning

- Configuring a search peer

- Configuring a search head

- Search head clustering

- Multisite indexer clustering

- Designing Splunk architecture

© Carlos Moreno Buitrago, Deep Mehta 2026

C. M. Buitrago and D. Mehta, *The Splunk Core User Study Companion*, Certification Study Companion Series, https://doi.org/10.1007/979-8-8688-2501-9_13

By the end of this chapter, you will have learned about designing architecture and resource planning in Splunk and be able to implement index clustering and search head clustering. By the time you complete this chapter, you will have covered 25% of the Splunk Enterprise Certified Architect exam blueprint.

Capacity Planning for Splunk Enterprise

Capacity planning plays a crucial role in scaling Splunk Enterprise. The components for capacity planning in Splunk Enterprise are as follows:

- An **indexer** is an instance that indexes data. The indexer provides data processing and storage for remote and local data as a peer node.

- A **search head** handles search management functions. It redirects search requests to peers and merges the results once the peer node processes the request.

- A **forwarder** is an instance that forwards data to indexers for data processing and storage.

Dimensions of a Splunk Enterprise Deployment

Capacity planning in Splunk Enterprise deployment has various dimensions that affect it:

- **Incoming data**: The greater the amount of data you send to Splunk, the greater the time and number of resources needed to process the incoming data.

- **Indexed data**: As the data in Splunk Enterprise increases, you need more I/O bandwidth to store data.

- **Concurrent users**: As more users use a Splunk instance at the same time, the instances require more resources to perform searches and create reports and dashboards.

- **Saved searches**: Splunk needs the capacity to run saved searches efficiently.

- **Premium Splunk products**, such as Splunk Enterprise Security, require more resources in the Splunk environment.

You start with the incoming data's effect on capacity planning.

Incoming Data Affects Splunk Enterprise Performance

Incoming data plays a critical role in capacity planning in Splunk. If there is a lot of data going to the indexer, but the physical memory is not large enough to process this data, indexer performance slows down. As data in Splunk increases, more system memory is utilized. Therefore, the amount of incoming data needs to be considered during capacity planning.

Indexed Data Affects Splunk Enterprise Performance

Indexed data also plays a critical role in capacity planning. Splunk Enterprise captures incoming data and gives it to the indexer. As the index increases, the available space decreases, and as the indexed data increases, searches slow down processing.

Concurrent Users Affect Splunk Enterprise Performance

Concurrent users greatly affect Splunk capacity planning. An indexer and Search Head need to provide a CPU core for every search that the users invoke. When multiple users are running searches, all the available CPU cores are quickly exhausted.

Saved Searches on Splunk Enterprise Performance

Concurrent users affect Splunk capacity planning. Saved searches, on average, consume about 1 CPU core and a fixed amount of memory. A saved search increases I/O when the disk system looks in the indexer to fetch data. If you schedule too many simultaneous saved searches, it leads to an exhaustion of resources.

Disk Storage for Splunk Enterprise

Splunk calculates disk storage based on the following formula:

```
(Daily average indexing rate) × (retention policy) × 1/2
```

- **Daily average indexing rate** stats incoming data in Splunk Enterprise.

- A **retention policy** is the parameter set in indexes.conf called FrozenTimePeriodInSecs.

- The compression ratio is usually 50%, and it will depend on the data type and the indexed fields we created.

Ideally, Splunk Enterprise stores raw data at approximately half its original size using compression and a lexicographical table. If Splunk Enterprise contains 150 GB of usable disk space, you can store nearly a hundred days' worth of data at an indexing rate of 3 GB/day.

Let's now move forward to learn about the three processes of configuring the search head and search peer.

Configuring a Search Peer

A Splunk Enterprise instance can run both as a search head and a search peer, called a standalone instance. In distributed environments, a search peer (indexers) can perform indexing and respond to incoming search requests from search heads.

Let's first discuss the process of configuring a search peer using Splunk Web.

Configuring a Search Peer in a Search Head from Splunk Web

To configure a search peer in a search head using Splunk Web, refer to the following steps:

1. In the search head, go to settings using Splunk Web, and go to Distributed Search.

2. Move to Search peers and click Add New.

3. Enter the IP address or the hostname of the indexer and enter **mgmt port-no** followed by a colon (*hostname:mgmt_port, IPaddress:mgmt_port*) for authentication to share a public key between the search head and indexer. Please provide the username and password of the remote indexer with admin permissions on the search head. (Refer to Figure 13-1.)

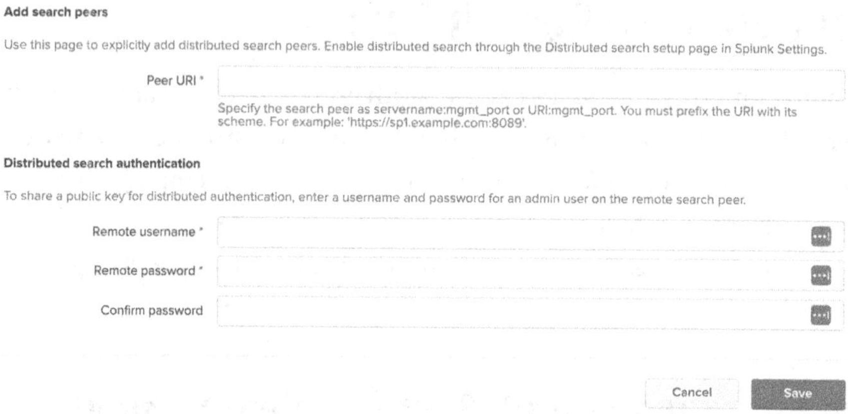

Add search peers

Use this page to explicitly add distributed search peers. Enable distributed search through the Distributed search setup page in Splunk Settings.

Peer URI *

Specify the search peer as servername:mgmt_port or URI:mgmt_port. You must prefix the URI with its scheme. For example: 'https://sp1.example.com:8089'.

Distributed search authentication

To share a public key for distributed authentication, enter a username and password for an admin user on the remote search peer.

Remote username *

Remote password *

Confirm password

Cancel   Save

Figure 13-1. Configure search peer

Configure Splunk Search Peer from the .conf File

To configure a Splunk search peer by editing the .conf file, refer to the following steps:

1. In the search head, create or edit distsearch.conf located in $SPLUNK_HOME/etc/system/local.

2. Add the following stanza in distsearch.conf:

```
[distributedSearch]
servers = <IPaddress1>:<mgmt_port>,<IPaddress2>:<mgmt_
port>,<IPaddress3>:<mgmt_port3>,.........<IPaddressn>:
<mgmt_port>
```

- **servers** are the addresses of the distributed search group server.

If you add a search head by editing .conf files, you need to distribute keys manually, as follows:

1. Copy the keys of the search head file from $SPLUNK_
   HOME/etc/auth/distServerKeys/trusted.pem
   to indexer location $SPLUNK_HOME/etc/auth/
   distServerKeys/<searchhead_name>/trusted.pem.

2. The <searchhead_name> is the search head
   serverName, which is configured in server.conf of
   the search head.

3. Restart the Splunk instance.

Configure Search Peer from Splunk CLI

To configure a search peer using Splunk CLI, refer to the following steps:

```
 splunk add search-server <IPaddress:mgmt_port or
hostname:mgmt_port> -auth <user>:<password> -remoteUsername
<username> -remotePassword <password>
```

- **search-server** is the address of the remote server.

- **-auth** is the username and password of the current
  instance.

- **-remoteUsername** is the username of the remote
  instance used for authentication.

- **-remotePassword** is the password of a remote instance
  used for authentication.

You have now covered the entire process of configuration methods used for the search peer.

The next section discusses search head clustering.

Configure a Search Head

In a distributed environment, a search head can perform searching and should not index data. The instances that manage search management functions generally direct search requests to the search peers. The search head directly communicates with the indexers.

Figure 13-2 shows the constituent structure of a Splunk search head.

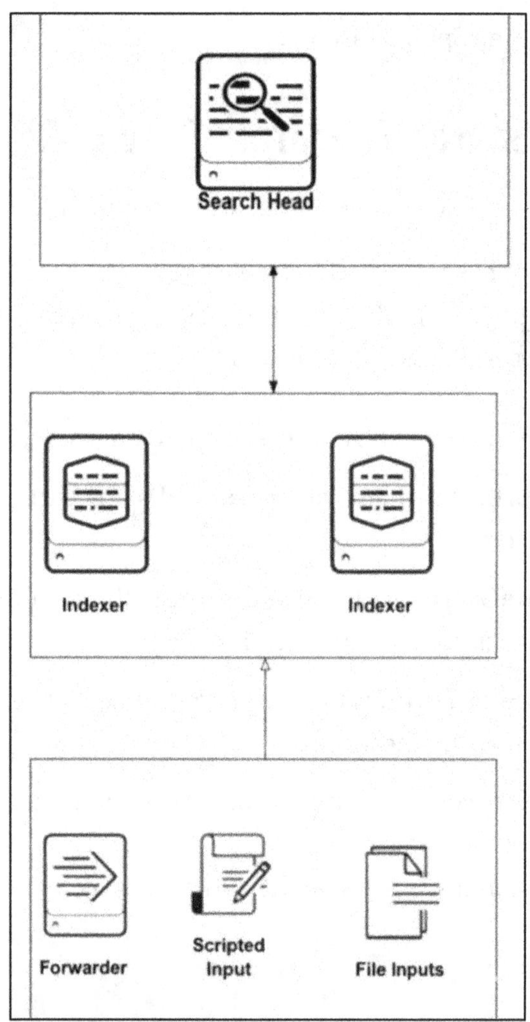

Figure 13-2. *Splunk architecture example*

The CPU and RAM need to be sized to allow the number of simultaneous searches (ad hoc and scheduled) on a single search head because ad hoc searches consume one core and increase I/O per execution. It is important to size the specs correctly since it could lead to skipped searches, searches that won't run, and missed alerts.

Let's now discuss the process of configuring a search head using Splunk Web.

Add a Search Head to an Index Cluster Using Splunk Web

You configure and enable the search head at the same time that you enable the other cluster nodes. The cluster's set of peer nodes becomes search peers of the search head. Refer to the following steps:

1. Go to Settings, and then go to the Distributed Search panel.

2. Move to Index Clustering, and click "Enable index clustering" as a Search Head Node, and click Next.

3. Enter the cluster manager IP address or the hostname. Enter **mgmt port number** followed by a colon (*hostname:mgmt_port, IPaddress:mgmt_port*) for authentication to share a public key between the search head and indexer. Please provide the common secret key of the manager node (see Figure 13-3).

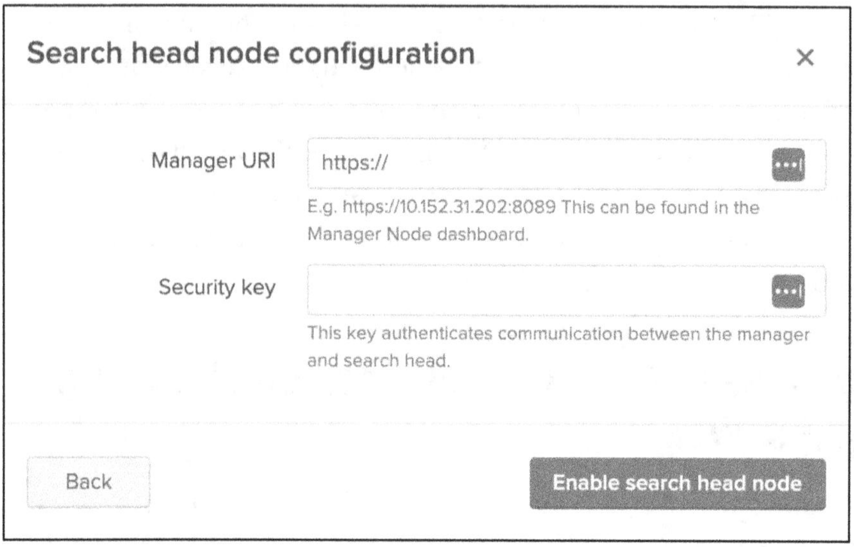

Search head node configuration   ×

Manager URI   https://

E.g. https://10.152.31.202:8089 This can be found in the Manager Node dashboard.

Security key

This key authenticates communication between the manager and search head.

Back   Enable search head node

Figure 13-3. *Configure search head*

Add a Search Head to an Index Cluster Using .conf file

To configure the search head clustering by editing the .conf file directly, refer to the following steps:

1. In the search head in Splunk, create or edit the server.conf located in $SPLUNK_HOME/etc/local.

2. Edit the following stanza in the server.conf files. Refer to the following stanza:

```
[clustering]
manager_uri =(IPaddress:mgmt_port or
hostname:mgmt_port)
mode = searchhead
pass4SymmKey = <key>
```

- **manager_uri** is the address of the manager node.

- **mode** is the type of clustering mode.

- **pass4SymmKey** is the cluster key.

After updating the configuration file, restart the Splunk instance.

Configuring a Search Head from Splunk CLI

To configure a search head using the Splunk CLI, use the following:

```
splunk edit cluster-config -mode searchhead -manager_
uri<IPaddress:mgmt_port or hostname:mgmt_port> -secret <key>
splunk restart
```

- **-manager_uri** is the address of the manager node.

- **-secret** is the cluster secret key.

Search Head Clustering

A search head cluster is a group of search heads that act as one. Members
share configurations, schedule searches, and replicate search artifacts
(knowledge bundles, dispatch artifacts) so the workload is balanced and
users see a consistent experience. A separate instance, called deployer,
pushes baseline app/config updates to all members. Within the cluster, an
elected captain coordinates job scheduling, artifact replication, member
health, and rolling maintenance. There are two types of captains:

- **Dynamic captain (normal mode)**: The captain is
  elected automatically from the active members. If the
  current captain fails or is taken down for maintenance,
  the cluster re-elects another member. This is the
  default and preferred behavior.

- **Static captain (forced/temporary):** In special
  situations, such as initial bootstrap, disaster recovery,
  or when the cluster cannot converge on a captain,
  you can force a specific member to be captain. This is
  temporary and mainly a recovery/bootstrapping tool;
  once the cluster stabilizes, it should return to dynamic
  election.

The search head clustering architecture is shown in Figure 13-4.

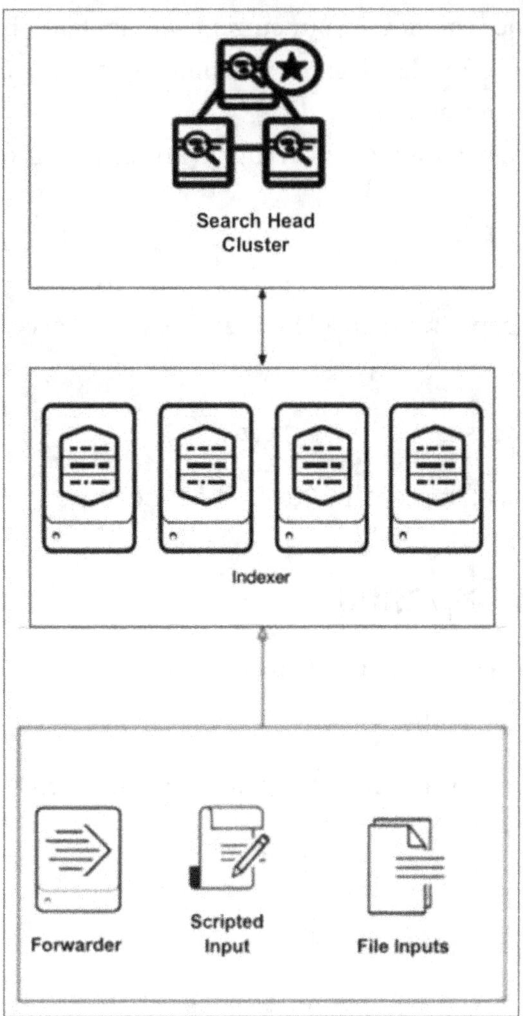

Figure 13-4. *Search head clustering architecture*

Search Head Cluster Captain

The captain is the search head cluster member that performs extra functions apart from search activities. It coordinates the activity of the cluster. You are able to make your search head run either scheduled searches or ad hoc searches. If your current captain fails, then a new captain is selected after the election. The new captain can be any Splunk search head cluster member.

To configure cluster members to run ad hoc searches, edit server.conf.

```
[shclustering]
adhoc_searchhead = true
```

To configure captain to run ad hoc searches, edit server.conf.

```
[shclustering]
captain_is_adhoc_searchhead = true
```

Let's now look at the captains' roles and functions.

The Role of Captains

The following are a captain's role in a Splunk search head cluster:

- Scheduling jobs

- Coordinating alerts and alert suppressions across the cluster

- Pushing the knowledge bundle to search peers

- Coordinating artifact replication

- Replicating configuration updates

Captain Election

A search head cluster uses a dynamic captain. When a condition arises, the cluster conducts an election, and any Splunk search head member can become a Splunk head captain. Captain election happens only when

- The current captain fails.

- There are network issues.

- The current captain steps down because it is not able to detect that major members are participating in the cluster.

- Network partition occurs.

Configure Search Head Cluster Using CLI in Splunk

To configure the search head clustering from Splunk CLI, do the following on all the search heads:

```
splunk init shcluster-config -auth <username>:<password> -mgmt_
uri <URI>:<management_port> -replication_port <replication_
port> -replication_factor <n> -conf_deploy_fetch_url
<URL>:<management_port> -secret <security_key> -shcluster_
label <label>
splunk restart
```

- **-auth** parameter specifies the current login credentials.

- **-mgmt_uri** specifies the URI of the current instance.

- **-replication_factor** parameter determines the number of copies of each search artifact that the cluster maintains. All cluster members must use the same replication factor. This parameter is optional. If not explicitly set, the replication factor defaults to 3.

- **-replication_port** specifies the port that the instance uses to listen to other artifacts. It can be any unused port.

- **-conf_deploy_fetch_url** parameter specifies the URL and management port for the deployer instance. This parameter is optional during initialization, but you do need to set it before you can use the deployer functionality.

- **-shcluster_label** is an option used to label a cluster.

- **-secret** is the key that specifies the security key.

The final step is to add the instance to the cluster. You can run the `splunk add shcluster-member` command either on the new member or from any current member of the cluster. The command requires different parameters depending on where you run it from.

Configure Dynamic Search Captain Using Splunk CLI

To configure a dynamic search captain in Splunk CLI, do the following:

```
splunk bootstrap shcluster-captain -servers_list
"<URI>:<management_port>,<URI>:<management_port>,..." -auth
<username>:<password>
splunk restart
```

Configure Static Search Captain Using Splunk CLI

To configure static search captain using Splunk CLI, refer to the following steps:

- On the member that you want to designate as captain, run this CLI command:

  ```
  splunk edit shcluster-config -mode captain -captain_uri
  <URI>:<management_port> -election false
  splunk restart
  ```

- On each non-captain member, run this CLI command:

```
splunk edit shcluster-config -mode member
-captain_uri <URI>:<management_port>
-election false
splunk restart
```

Multisite Indexer Clustering

Multisite indexer clustering is an indexer cluster that consists of multiple sites. Each site needs to follow specific replication and search factor rules. Multisite clusters offer two key benefits over single-site clusters:

- **Improved disaster recovery**: By maintaining multiple copies of data at multiple locations, you save your data in a disaster. Multisite indexer clustering keeps multiple copies of data at multiple locations and provides site failover capabilities.

- **Search affinity**: This configures each site to have local data and a full set of searchable data. The search head on each site limits searches to peer nodes only. This eliminates any need, under normal conditions, for search heads to access data on other sites, greatly reducing network traffic between sites.

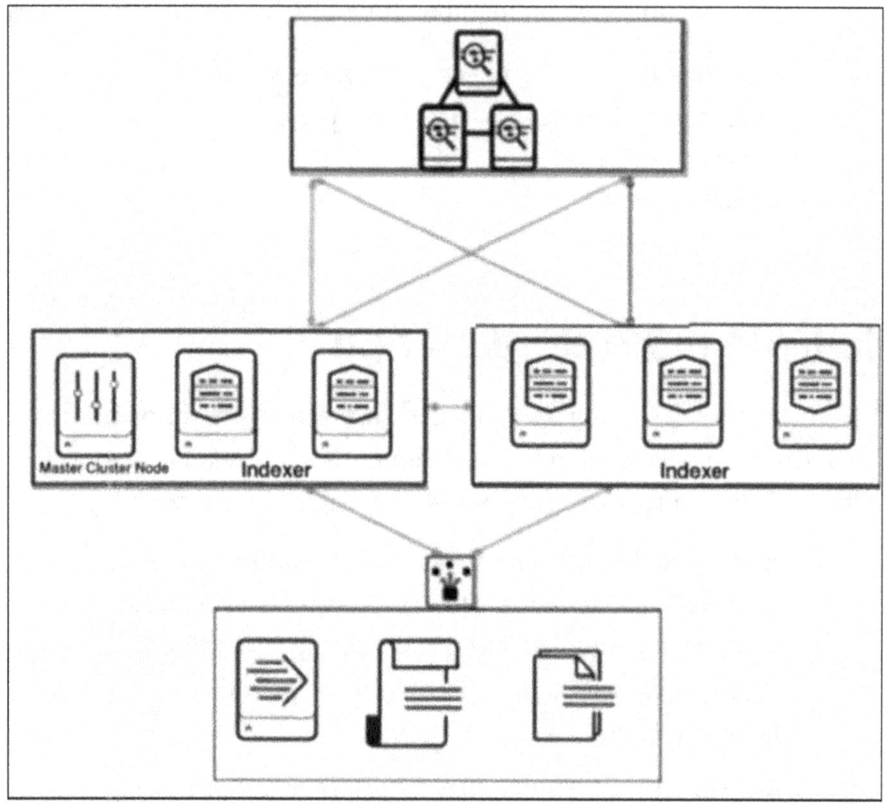

Figure 13-5. *Multisite index clustering architecture*

Configure Multisite Indexer Clustering Using .conf Files

To configure multisite indexer clustering by editing the .conf file, refer to the following steps:

1.   Create or edit the server.conf located in $SPLUNK_ HOME/etc/system/local, or deploy a custom app to manage the configuration.

2. Edit the following stanza in the server.conf files
    on all multisite cluster nodes (manager/peers/
    search heads):

```
[general]
site = site1,site2,site3,.......,site n
```

3. On the manager node and search heads:

```
[clustering]
multisite = true
```

4. On the manager node only:

```
[clustering]
mode = manager
available_sites = site1,site2,......,site n
site_replication_factor =
origin:<number>,total:<number>
site_search_factor =
origin:<number>,total:<number>
pass4SymmKey = <key>
cluster_label = <cluster name>
```

5. On the peer nodes:

```
[clustering]
mode = peer
manager_uri = <manager_URI>
pass4SymmKey = <key>
```

6. On the search heads:

```
[clustering]
mode = searchhead
manager_uri = <manger_URI>
pass4SymmKey = <key>
```

- **available_sites** is used in a manager node to show which sites are available for the Splunk environment.

- **site_replication_factor** specifies the total copies of raw data that the cluster should maintain.

- **site_search_factor** is used in a manager node to maintain several searchable copies in the Splunk environment for disaster recovery.

- **origin:<n>** specifies the minimum number of copies of a bucket that will be held on the site originating the data in that bucket (i.e., the site where the data first entered the cluster). When a site originates the data, it is known as the "origin" site.

- **site1:<n>, site2:<n>, ...,** indicates the minimum number of copies that will be held at each specified site. The identifiers "site1", "site2", and so on are the same as the site attribute values specified on the peer nodes.

- **total:<n>** specifies the total number of copies of each bucket, across all sites in the cluster.

- **pass4Symmkey** authenticates communication between nodes.

- **cluster_label** is the name of the cluster.

Configure Splunk Multisite Indexer Clustering Using CLI

To configure a Splunk instance as a manager node for a multisite indexing cluster in Splunk using CLI commands, refer to the following commands:

```
splunk edit cluster-config -mode manager -multisite true
-available_sites site1,site2 -site site1 -site_replication_
factor origin:2,total:3 -site_search_factor origin:1,total:2
-secret your_key
splunk restart
```

- **available_sites** is used in a manager node to show which sites are available for the Splunk environment.

- **site_replication_factor** specifies the total copies of raw data that the cluster should maintain.

- **site_search_factor** is used in a manager node to maintain several searchable copies in the Splunk environment for disaster recovery.

- **origin:<n>** specifies the minimum number of copies of a bucket that will be held on the site originating the data in that bucket (i.e., the site where the data first entered the cluster). When a site originates the data, it is known as the "origin" site.

- **site1:<n>, site2:<n>**, ..., indicates the minimum number of copies that will be held at each specified site. The identifiers "site1", "site2", and so on are the same as the site attribute values specified on the peer nodes.

- **total:<n>** specifies the total number of copies of each bucket, across all sites in the cluster.

- **pass4Symmkey** authenticates communication between nodes.

- **cluster_label** is the name of the cluster.

To configure a peer node in a multisite cluster using CLI commands, refer to the following commands:

```
splunk edit cluster-config -mode peer -site <n> -manager_uri
<ip>:<port>/<hostname>:<port> -replication_port <port number>
-secret <key>
splunk restart
```

To configure a search head for a multisite indexing cluster using CLI commands, refer to the following commands:

```
splunk edit cluster-config -mode searchhead -site <n> -manager_
uri <ip>:<port>/<hostname>:<port> -secret <key>
splunk restart
```

Before moving forward to discuss the design of the Splunk architecture, let's review what you have already learned in this chapter: capacity planning in the Splunk environment, configuring search heads, search head clustering, captains, and clustering the multisite indexer. The next section discusses designing and architecture in Splunk.

Designing Splunk architecture is a complex job. To design in Splunk, you should be aware of the series of operations in its architecture and all the software components.

- **Data input**: To input data in Splunk, you use universal forwarders or heavy forwarders and forward it to the indexers.

- **Data parsing**: To parse data, you use heavy forwarders and the indexers.

- **Data indexing**: To index data in a Splunk environment, you use the indexers.

- **Data searching**: To search data in a Splunk environment, you use search heads.

At this point, you have already revised the Splunk architecture with all the required components for data input, data parsing, data indexing, and searching data. Implementing Splunk architecture is not a big issue based on the diagram. But how do you design a Splunk architecture?

To understand how to design, you need to know about Splunk Validated Architectures.

Splunk Validated Architectures (SVAs)

Splunk Validated Architectures (SVAs) were adapted to achieve a stable, efficient, and repeatable deployment. SVAs ensure platform scaling and troubleshooting on the Splunk platform and reduce the total cost of ownership.

The following are the characteristics of SVAs:

- **Performance**: SVAs help organizations improve performance and help with stability.

- **Complexity**: SVAs help organizations remove complexity from their environment, as complexity acts as the biggest barrier when planning for scaling Splunk environment.

- **Efficiency**: SVAs help organizations achieve efficiency by improving operations and accelerating time to value.

- **Cost**: SVAs help organizations reduce the cost of the organization by reducing the cost of ownership.

- **Agility**: SVAs help organizations adapt as they scale up when data grows.

- **Maintenance**: SVAs help organizations reduce the organization's maintenance efforts by channeling resources in the proper dimension.

- **Scalability**: SVAs help organizations scale efficiently and seamlessly.

- **Verification**: SVAs help organizations ensure that the Splunk deployment is built on best practices.

Designing Splunk Validated Architectures

When designing a Splunk architecture, you need to understand a few key points to maintain your workflow.

- **Availability**: The Splunk environment needs to be operational at any time and recover from planned and unplanned outages or disruptions.

- **Performance**: The Splunk environment needs to maintain the same level of service under different conditions.

- **Scalability**: The Splunk environment needs to be designed to scale on all tiers and handle the workload effectively.

- **Security**: The Splunk environment should protect data, configurations, and assets.

- **Manageability**: The Splunk environment is centrally operated and manages all the tiers.

Topology Categories

This section explains the SVA topology categories. You'll use these labels in the questionnaire, and you'll see them again in the next steps of the SVA selection process. Refer to Table 13-1 and Table 13-2.

Table 13-1. *Indexing tier categories*

Category Code	Explanation
S	It indicates the indexer of a single-server Splunk environment
D	It indicates the need for a distributed indexer tier
C	It indicates the need for a clustered indexer tier (data replication)
M	It indicates the need for a multisite clustered indexer tier

Table 13-2. *Search tier categories*

Category Code	Explanation
1	It indicates a single search head may meet the requirements
2	It indicates multiple search heads are required to meet requirements
3	It indicates a search head cluster is required to meet requirements
4	It indicates a search head cluster that spans multiple sites (a "stretched" SHC) is required to meet requirements
+10	It indicates a dedicated search head (cluster) is required to support Enterprise Security App. Add 10 to the search tier topology category, and carefully read the description for the topology for specific requirements for this app

Small-Scale Enterprise Deployment

If the indexing volume is less than 500 GB/day and the organization has a small number of users with non-critical search use cases and does not have any requirements to provide high availability or automated disaster recovery, a small-scale enterprise deployment is favorable. A small-scale enterprise deployment has various forwarders to forward the data to the indexer. The indexer and search capabilities are placed together to use the server more efficiently.

Figure 13-6 shows a small enterprise deployment.

- **Data input:** Data enters through forwarders, scripted input, file input, and so forth, according to the settings and deployment. The data is forwarded to the idx/sh.

- **Indexing/searching**: The instance in this type of
  deployment receives data from forwarders. It stores
  and indexes incoming data. It runs user queries
  and performs a local search on data from the data
  input phase.

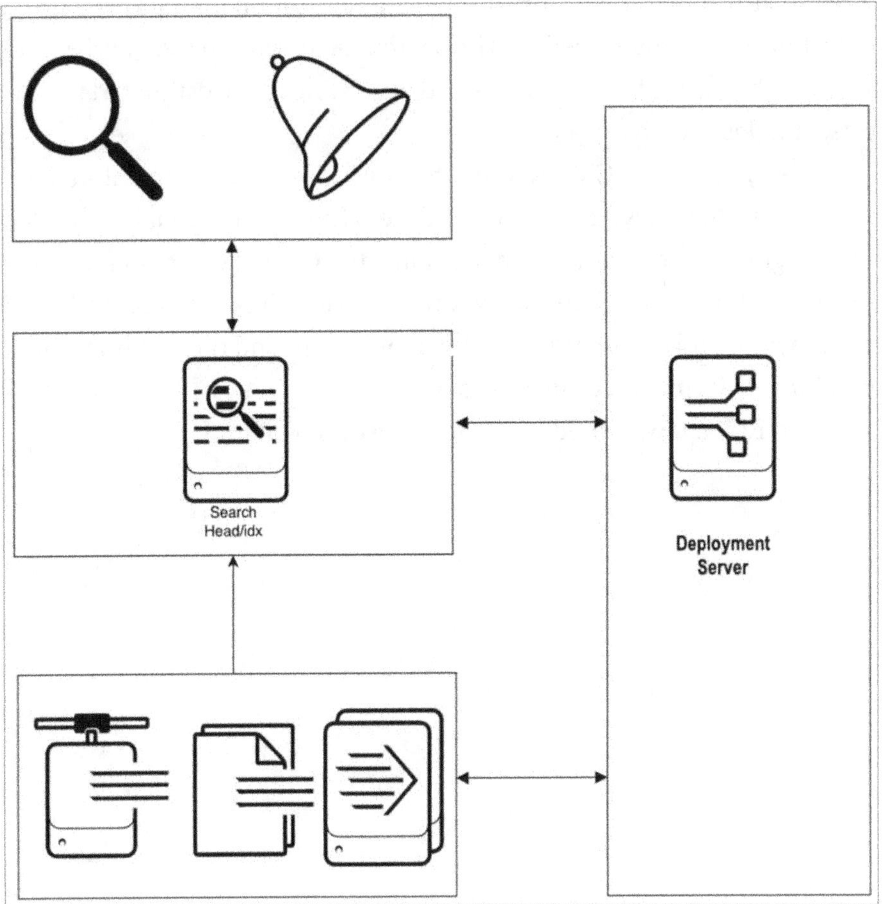

Figure 13-6. *Small enterprise Splunk constituent architecture deployment*

Medium-Scale Enterprise Deployment (C1/C11)

If the indexing volume exceeds the capacity of a single-server deployment, it requires scalable, highly available data ingest and data resiliency in the case of failure of a single indexer node; then a medium-scale enterprise deployment is the recommended one. It has various forwarders forwarding the data to the index cluster, providing data replication for backups and search improvements if needed. The search head will run the queries over the index cluster, and it should be sized according to the estimated users, alerts, and dashboards.

A cluster manager (CM) coordinates the indexer cluster to enforce the data-replication policy, track which indexers (peers) are available, and let you configure the search head once against the CM instead of managing peers individually. The Monitoring Console (MC) gives you a central view of health and capacity across the deployment and raises alerts when something drifts into an unhealthy state.

Figure 13-7 shows a medium enterprise deployment.

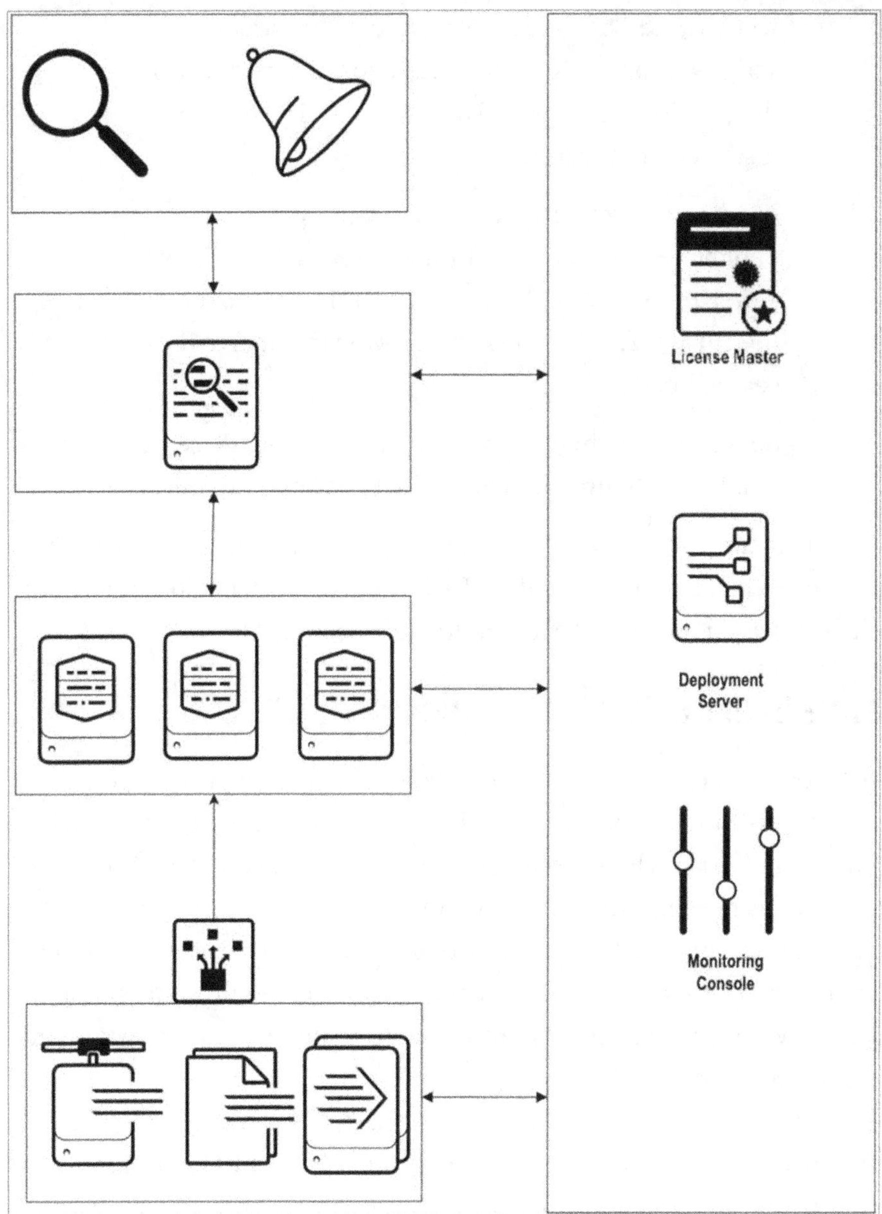

Figure 13-7. *Medium enterprise Splunk constituent architecture deployment*

- **Data input**: The data enters through forwarders, scripted inputs, file input, and so forth, according to the required settings and deployment. The data is forwarded to the index cluster.

- **Indexing**: Indexers in this type of enterprise deployment receive data from the forwarders. They store, index, and replicate the incoming data over the cluster. The indexers will provide the data to the search head.

- **Searching**: In this type of enterprise, a separate search head is configured to handle the users' query requests.

If the incoming data suddenly increases in a medium-scale enterprise, you need to make it more scalable. To do so, you need to add more indexers, search heads and even multisite if needed.

Large-Scale Enterprise Deployment (C3/C13)

The distributed clustered deployment with SHC in a single site uses clustering to add horizontal scalability and remove the single point of failure from the search tier by using a search head cluster (SHC). The SHC deployer has no runtime/HA role; it's only for pushing app/config updates. Additionally, it is important to know that a minimum of three search heads is required to form an SHC, and the topology needs a third-party load balancer with sticky sessions in front of SHC members so a user stays on the same SH during a session.

The topology provides more search capacity than a single SH, the scheduled searches will be spread across the cluster, and failover tolerance in the search layer. Figure 13-8 shows a large enterprise deployment.

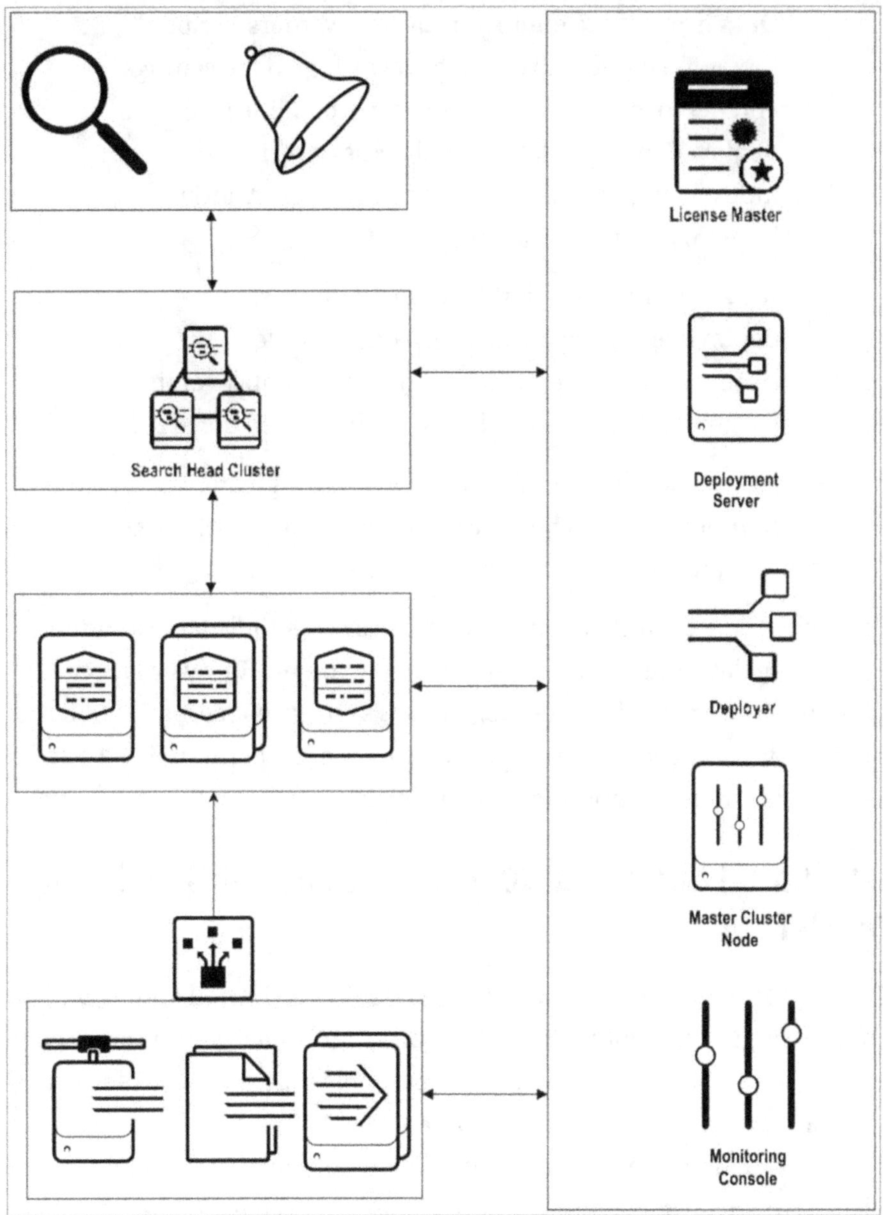

Figure 13-8. *Large enterprise Splunk constituent architecture deployment*

- **Data input**: Data enters through forwarders, scripted input, file input, and so forth, according to the settings and deployment. Load-balancing capabilities are deployed to make data spread across a series of indexers in case it's needed. The Splunk forwarders have this feature by default.

- **Indexing**: Indexers in this type of enterprise deployment receive data from the forwarders. They store, index, and replicate the incoming data over the cluster. The indexers will provide the data to the SHC.

- **Search head**: In this type of enterprise, a greater number of search heads are configured to handle large amounts of query requests from users.

By now, you know that Splunk Enterprise can handle large amounts of incoming data, but from a single-site deployment. However, since all components are placed in one site, it raises security concerns. What if the single site is damaged in a disaster, like a flood or an earthquake? To avoid this situation, Splunk supports multisite clustering.

Distributed Clustered Deployment with SHC Multisite (M3/M13)

Multisite (M3/M13) uses search head cluster for horizontal scaling and removes single point failure from the search tier in each site. It has a search head cluster (SHC) per site to increase available search capacity, distribute scheduled workload, and provide optimal user failover. Multisite (M3/M13) also includes the implementation of distributed cluster deployment, useful to replicate data among multiple sites by configuring site replication and search factor.

Multisite indexer clustering is depicted in Figure 13-9.

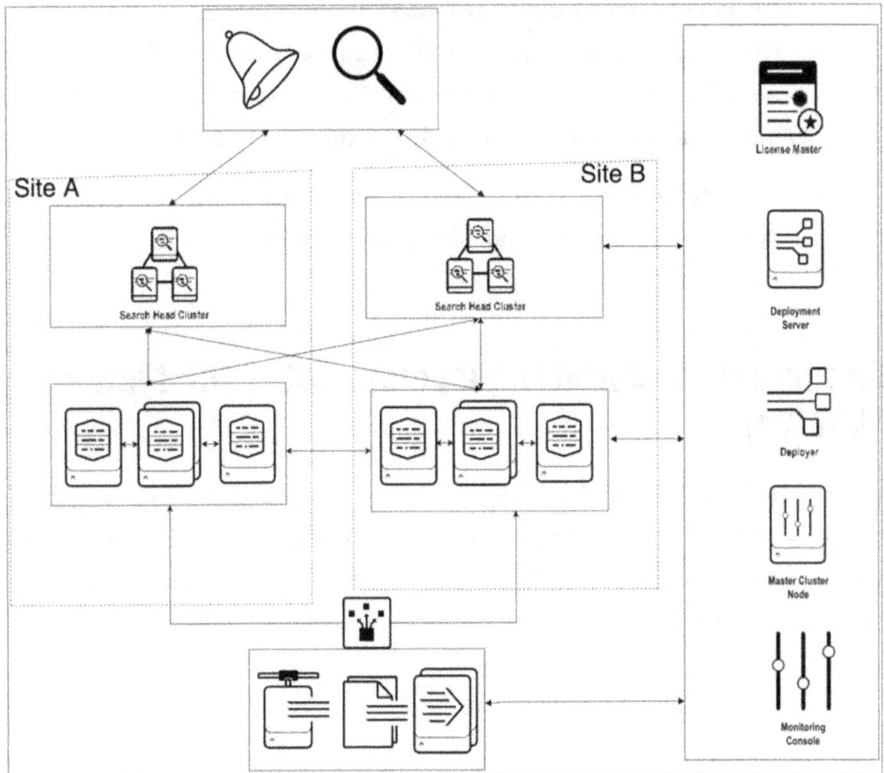

Figure 13-9. Multisite (M3/M13) constituent architecture deployment

- **Data input**: Data enters through forwarders, scripted input, file input, and so forth, according to the settings and deployment. Load-balancing capabilities are deployed to make data spread across a series of indexers in case it's needed. The Splunk forwarders have this feature by default.

- **Indexing**: Indexers in this type of enterprise deployment receive data from the forwarders. They store, index, and replicate the incoming data over the cluster. The indexer also has a manager node that regulates the function of the indexer cluster and coordinates how buckets are replicated among different sites. The indexers will provide the data to the SHC.

- **Search head clustering**: Search head clusters are configured for individual sites. Each site has its own search head clustering.

Distributed Clustered Deployment with SHC: Multisite (M4/M14)

Multisite (M3/M13) and Multisite (M4/M14) are almost similar. The only change is that it includes one stretched search head cluster that manages two sites. It results in optimal failover for users in case of a search node or data center failure.

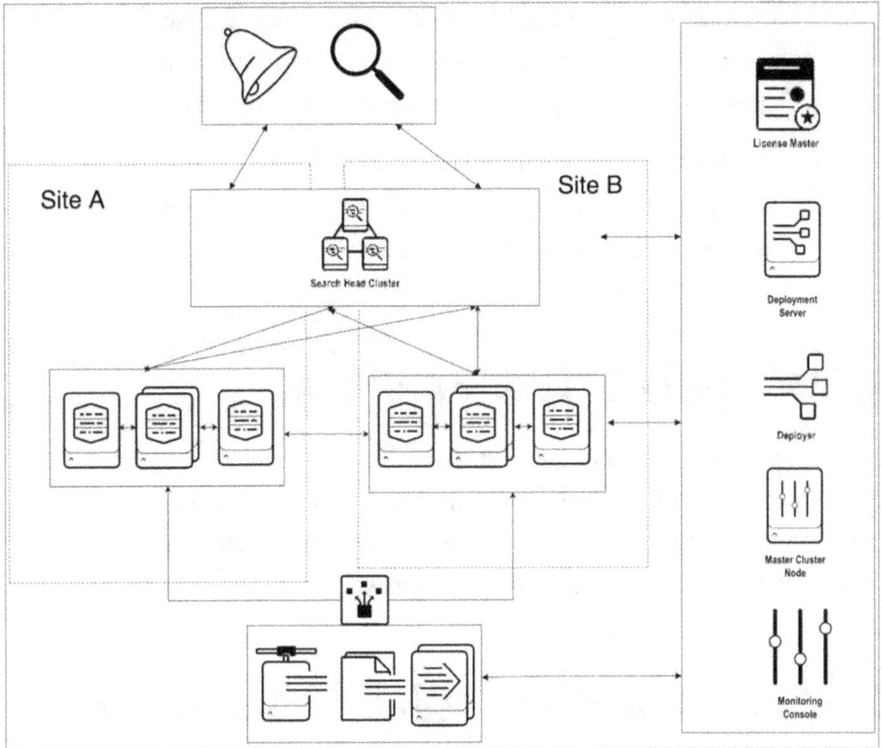

Figure 13-10. Multisite (M4/M14) constituent architecture deployment

- **Data input**: Data enters through forwarders, scripted input, file input, and so forth, according to the settings and deployment. Load-balancing capabilities are deployed to make data spread across a series of indexers in case it's needed. The Splunk forwarders have this feature by default.

- **Indexing**: Indexers in this type of enterprise deployment receive data from the forwarders. They store, index, and replicate the incoming data over the cluster. The indexer also has a manager node that regulates the function of the indexer cluster and coordinates how buckets are replicated among different sites. The indexers will provide the data to the SHC.

- **Search head clustering**: Search head clusters are configured according to your enterprise design. Search head clustering is stretched across shared sites.

Splunk Architecture Practices

This section offers a peek at the Splunk architecture roles and the practices. There is one problem statement. A company named ACME Corporation wants to move to Splunk. ACME has provided its requirements, so let's use them to apply best practices in migrating existing solutions to Splunk.

Use Case: Company ACME Corporation

Your client, ACME Corporation, wants to move to Splunk. It instituted its current infrastructure to meet PCI compliance ecommerce requirements. The logging system implemented to address those requirements is now at overflow. ACME is running out of resources to store logs, which is affecting their ability to monitor their operational, security, and compliance logs. As a result, ACME wants to move to Splunk. Its current environment uses Linux and Windows-based logs. It directly correlates with syslog.

The current data sources are firewall, ecommerce services, proxy, database, and network logs. The retention period differs among the different data sources. For firewalls and proxy data, the retention period is 90 days. For app services, the retention time is 120 days. For ecommerce, the retention time is 365 days. For databases, the retention is 180 days. For networks, the retention is 30 days. App services, networks, and proxies are Windows and Linux-based. The others are Linux only.

PCI compliance is only for ecommerce data; the others do not have a compliance-based data feed. The level of access is defined in four levels: security, Ops, sales, and support. Security has access to all logs. The Ops team has access to the app services, ecommerce, database, and network logs. The support team has access to app services and ecommerce logs. The sales team has access to only the ecommerce logs.

The raw data compressed will take around 15% of the original size (provided), and the compressed index fields will take around 35% of the original size (provided). The average amount for a firewall data source is 40 GB/day. App services have 20 GB/day, ecommerce has 102 GB/day, a proxy data source has 30 GB/day, databases have 40 GB/day, and the network has 10 GB/day.

After discussions with the client, you understand how you can monitor the firewall, ecommerce, proxy, and database data sources. For app services, you need to use scripted input. To capture network logs, you need to use network ports.

Company ACME has 38 users. It has 9 users in the security group; they are active from 12 to 18 hours/day. It has 12 users in the sales group; they are active from 7 to 11 hours/day. It has 10 users in the Ops group; they are active 6 to 13 hours/day. It has 9 users in the support group; they are active 2 to 8 hours a day.

The company needs 40 dashboards, 75 alerts, 85 reports, and 5 data models. The data rolling period from a hot bucket to a cold bucket is 30 days.

The Splunk architecture is designed to have six stages.

- **Splunk data input** is the initial stage of data gathering. Solution architects had a series of calls with company ACME to gather requirements such as daily incoming data, data sources, retention periods, and access visibility to each group.

- **Splunk index calculation** is the second stage. It is based on the raw data compression ratio, retention days, and the index file compression ratio. You calculate the base size of raw data and the base size of index calculation, which helps you design the Splunk index and get an overview of the current architecture.

- **Splunk total saved data** is the third stage. In this stage, you calculate the total required disk size based on the raw data compression ratio, retention days, index file compression ratio, replication factor, and search factor. Total saved data provides insight to the Splunk architect to help him set the number of indexers needed for the infrastructure.

- **Splunk user planner** is the fourth stage. The overhead cost is calculated based on scheduled searches, concurrent users, active time of concurrent users, and index parallelization. The Splunk user planner helps the architect determine the overhead costs and plan the Splunk infrastructure design.

- **Splunk hardware and apps considerations** are the fifth stage. Disk storage, horizontal scaling, vertical scaling, index parallelization, and sizing factors for searching are investigated in this stage. Hardware and Splunk scaling considerations help Splunk architects design a scalable solution.

- **Disk size calculation** is the last stage. In this stage, you calculate the exact disk size based on all previous stages' data feed. In disk size calculation, you learn how Splunk architecture is designed.

Now let's dive into each stage.

Splunk Data Inputs

Splunk data input is the initial stage. It comprises gathering the data sources, methods, environment, index names, retention days, visibility access, ownership, integrity, raw data compression ratio, index file compression ratio, and average daily amount of data in gigabytes. Let's have a look at each component.

- **Data Source** is the name of the data input or data source.

- **The method** describes where the data is located and how to onboard data to Splunk. It can be collected through scripted input, monitor stanzas, network ports, and so forth.

- **The environment** is the operating system. It helps in planning the enterprise deployment.

- **Index Name** is the name of the index that the Splunk architect needs to plan. It helps in effective deployment.

- **Retention Days** is the number of days that data should be kept in Splunk. A retention policy depends on legal, business, and search requirements.

- **Visibility/Access** specifies which individuals or teams have access to data.

- **Ownership** defines the owner or person who controls Visibility/Access and can make notable changes in Splunk.

- **Integrity** specifies special considerations needed to maintain data integrity.

- **Compression Raw Data** is an estimated percentage of what Splunk saves by compressing raw data. For example, 15% means 100 GB of raw data can be compressed to 15 GB.

- **Compression Index File** is an estimated ratio of what Splunk saves by compressing the index file. For example, if there is 100 GB of incoming raw data, and the compression index file (lexical table) will take 35%, the incoming data is compressed to 35 GB.

- **Average Data GB per Day**: The daily average amount of incoming data in gigabytes.

Figure 13-11 is a Splunk data input spreadsheet. Let's have a look at it.

Data Source (Name)	Method (Name)	Environment (linux&Windows)	Index Name (Index name)	Retention Days (In Days)	Visibility/Access (Group Name)	Ownership (Group Name)	Integrity (Yes/No)	Compression Raw Data (In %)	Compression Index File (In %)	Average Data GB Per Day (in Gigabit)
Firewall	Monitor	Linux	firewall	90	Security	Security	No	15%	35%	40
App Services	Scripted Input	Windows,linux	app	120	Security,Ops,Support	Security	No	15%	35%	20
Ecommerce	Monitor	linux	ecom	365	security ,Ops,Support,Sales	Security	Yes	15%	35%	102
Proxy	Monitor	linux,windows	proxy	90	Security	Security	No	15%	35%	30
Database	Monitor	linux	db	180	Security,Ops	Security	No	15%	35%	40
Network	Network Port	Windows,linux	network	30	Security,Ops	Security	No	15%	35%	10
										242

Figure 13-11. *Splunk data input*

Splunk Index Calculation

Splunk index calculation is the second stage. Let's look at each component.

- **Compression Raw Data** is an estimated ratio of what Splunk saves by compressing raw data. **Compression Raw Data = compression percentage * Average Data per Day**. For example, the firewall data source has 40 GB of average data per day, and the size of compressed data will be 15%. So, raw data compression for a firewall source type is 6 GB/day.

- **Compression Index File** is an estimated ratio of what Splunk saves by compressing the index file. **Compression Index File = compression index file percentage * Average Data per Day.** For example, the firewall data source has 40 GB of average data per day, and the size of compressed index data will be 35%. So, the compression index file for a firewall source type is 14 GB/day.

- **Base Size of Raw Data**: The raw data's base size constitutes the total size of raw data per data source. To calculate, **Base Size of Raw = Compression Raw Data * Retention Days.** For example, the firewall data source has 6 GB/day for raw data compression and retention is 90 days. So, the base size of raw data for the firewall source type is 540 GB.

- **Base Size of Index File**: The base size of raw data constitutes the total raw data per index file. To calculate, **Base Size of Index=Compression Index File * Retention Days.** For example, the firewall data source has 14 GB/day for data index and retention is 90 days. So, the base size of index data for the firewall source type is 1260 GB.

Figure 13-12 is a Splunk index calculation spreadsheet. Let's have a look at it.

Data Source	Method	Environment	Index Name	Retention Days	Visibility / Access	Ownership	Integrity	Compression Raw Data	Compression Index File	Average Data Per Day GB
Firewall	Monitor	Linux/Windows	firewall	90	Security/Ops	Security	Yes	15%	35%	120
App Services	Scripted	Windows/Linux	app	120	Security/Ops	Security	No	15%	35%	50
Proxy	Monitor	Linux/Windows	proxy	60	Security/Ops	Security	Yes	15%	35%	80
Database	Monitor	Linux	db	180	Security/Ops	Security	No	15%	35%	40
Network	Network Port	Windows/Linux	network	30	Security/Ops	Security	No	15%	35%	70

Figure 13-12. *Splunk index calculation*

Splunk Total Disk Size

Splunk Total Disk Size is the third stage. Let's look at each component.

- **Search Factor** specifies the number of searchable buckets inside index clustering. Searchable buckets have both raw data and index files. When you configure a manager node in Splunk Enterprise, you set a search factor that defines the number of searchable copies per bucket that must be maintained in the entire index cluster. The default value of the search factor is 2.

- **Replication Factor** specifies the total number of raw data copies the cluster should maintain. Indexers store incoming data in buckets, and the cluster maintains copies of each bucket across the environment. Generally, the replication factor value is one more than the search factor.

- **Total Disk Size** comprises the base size of raw data, the base size of the index file, the search factor, and the replication factor. To calculate, Total Disk Size = Base Size of Raw Data * Replication Factor + Base Size of Index File * Search Factor. For example, when you want to calculate total disk size, you consider that the firewall data source has a Base Size of Raw Data = 540 GB, Base Size of Index File = 1260 GB, Search Factor = 2, and Replication Factor = 3. You get Total Disk Size = 4140 GB.

Figure 13-13 is a Splunk total disk size spreadsheet based on information in the use case. Let's have a look at it.

Data Source	Retention Days	Base Size Of Raw Data	Base Size Of Index file	Search Factor	Replication Factor	Total Disk Size
(Name)	(In Days)	(In GB)	(In GB)	(Count)	(Count)	(In GB)
Firewall	90	540	1260	2	3	4140
App Services	120	360	840	2	3	2760
Ecommerce	365	5584.5	13030.5	2	3	42814.5
Proxy	90	405	945	2	3	3105
Database	180	1080	2520	2	3	8280
Network	30	45	105	2	3	345
		8014.5	18700.5			61444.5

Figure 13-13. *Splunk total disk size*

Splunk User Planner

The Splunk User Planner is the fourth stage. Let's look at each component.

- **Ad hoc searches** are also called unscheduled searches. They can be run in several ways. For example, when you are invoking a search command using the Splunk search processing language, it invokes ad hoc searches.

- **Scheduled searches** are scheduled using the search scheduler or created as part of a report acceleration or data model acceleration. The report, dashboard, and alert are the **searches** that are generated for report acceleration or data model acceleration. Summarization reports are also a part of summarization searches.

- **Index parallelization** allows the indexer to maintain multiple pipeline sets. It allows the indexer to create and manage multiple pipelines, provide more cores to process them, and increase the limit of indexer I/O capacity.

- **Real-time searches** refresh in real time, continually pushing old data and acquiring new data, depending on the bounds.

- **Concurrent users** provide a CPU core for every search that the user invokes. If multiple users are logged in and running the searches, all the available CPU cores are quickly exhausted. For example, consider four cores and 12 users who are running different searches. The time to process search increases because the number of available cores is less than the number of requests.

Table 13-3 is a Splunk user planner summary based on information in the use case. Let's have a look at it.

Table 13-3. *Splunk user planner*

Parameter	Source Name	Count
Scheduled Searches	Alerts	75
	Reports	85
	Accelerate Data Model	5
Max Concurrent Users	Security	7
	Sales	12
	Ops	10
	Support	9

Hardware and Splunk Scaling Considerations

Hardware and Splunk scaling is the fifth stage. Let's look at each component.

- **Disk storage** comprises input/output operations per second. It helps measure disk storage. Hard drives read and write at different speeds. For bucket storage,

Splunk uses a storage area network (SAN) and network-attached storage (NAS). SAN is for hot and warm buckets, and NAS is used for a cold bucket.

- **Horizontal scaling**: If a load increases and you want to add more indexing power to the current Splunk architecture pool, you can use horizontal scaling to upgrade it by adding a greater number of indexers. Similarly, to provide more searching power, you can add more search heads to existing resources.

- **Vertical scaling**: If a load increases and you want to add more indexing power to the current Splunk architecture pool, you can use vertical scaling to upgrade the machine configuration by adding more CPU and cores. But there are limitations.

- **Increase index parallelization** is used only when the current system is underutilized. For example, you have 12 core indexers, but only four cores are being used; it is recommended to increase index parallelization.

- **Sizing factors for searching** comprise a number of scheduled searches, the number of reports, total dashboards, active concurrent users, total count of searches per hour, and so forth. On the basis of all such factors, sizing factors for searching need to be designed.

Disk Size Calculation

Disk size calculation is the sixth stage of index calculation. In this stage, you design a Splunk architecture index in which you calculate the disk size.

From the information presented in the use case, you see that the ingestion volume in Splunk is 242 GB/day. The replication factor is 3, and the search factor is 2. Hot and warm buckets are saved in SAN. Cold buckets are saved on NAS. According to the use case, data is in hot buckets and warm buckets for 30 days. You have four indexers with 60.5 GB volume of data indexing per day, and the overhead cost is 25%.

Figure 13-14 is a spreadsheet for disk size calculation using the information from the use case.

Data Sizing Exercise

Replication Factor	3											
Search Factor	2											
Number of Days in Hot/Warm	30											

Data Source	GB per day	Raw Compression Rate	Index Compression Rate	Retention in Days	Base Size of Raw	Base Size of Index Files	Replicate (Y or blank)	Replicated Size on Disk	Hot and Warm	Cold	Visible to	Index
firewall	40	0.15	0.35	90	540	1,260	Y	4,140	1,380	2,760	Security	firewall
app	20	0.15	0.35	120	360	840	Y	2,760	690	2,070	Security,Ops,Support	app
ecom	102	0.15	0.35	365	5,585	13,031	Y	42,817	3,520	39,297	security ,Ops,Support,Sales	ecom
proxy	30	0.15	0.35	90	405	945	Y	3,105	1,035	2,070	Security	proxy
db	40	0.15	0.35	180	1,080	2,520	Y	8,280	1,380	6,900	Security,Ops	db
network	10	0.15	0.35	30	45	105	Y	345	345	-	Security,Ops	network
Totals	242				8,015	18,701		61,447	8,350	53,097		

Number of Indexers	4	60.5	Ingestion Volume per Indexer GB/Day			
Recovery space	if number of indexers lost=	1		15,362	2,088	13,274
Overhead	25%			19,202	2,610	16,592
Total Disk				96,011	13,048	82,963
Disk Space per Indexer				24,003	3,262	20,741

Figure 13-14. *Disk size exercise*

Let's quickly recap this use case and what's needed to calculate the disk size.

- **Rep factor:** 3

- **Search factor:** 2

- **Number of indexers:** 4

- **Number of days in hot/warm bucket:** 30 days

- **Overhead**: 0.25 or 25%

- **Number of indexer lost**: 1

- **Raw compression rate:** 0.15

- **Index compression rate:** 0.35

- **Base size of raw**: Compression Raw Data * Retention Days

- **Base size of index files**: Compression Index File * Retention Days

- **Replicate size on disk**: Base Size of Raw Data* Replication Factor + Base Size of Index File * Search Factor

- **Ingestion volume per indexer**: Total GB per day/ Number of Indexers

- **Hot and warm**: (Raw Compression Rate * GB per Day * Replication Factor) + (Index Compression Rate * GB per Day * Search Factor) * Number of days in hot/ warm bucket

- **Cold bucket**: (Raw Compression Rate * GB per Day* Replication Factor) + (Index Compression Rate * GB per Day * Search Factor) * (Retention Days – Number of days in hot/warm bucket)

- **Data storage per indexer**: Total Replicated Size on Disk /Number of Indexer

- **Overhead cost**: Data Storage per Indexer + Overhead * Data Storage per Indexer

- **Total disk space**: Overhead Cost * Number of Indexers + Overhead cost

- **Total disk space per indexer**: Total Disk/Number of Indexer

With this, you have come to the end of this chapter. Congratulate yourself on successfully learning about capacity planning in the Splunk environment, single and multisite clustering, search heads, and designing a Splunk architecture in an enterprise deployment.

Summary

This chapter addressed various capacity planning components and configuring search heads using Splunk Web, the CLI, and the configuration file. It also covered search head clustering, the configuration of the search head captain, multisite clustering, and the Splunk architecture. You learned about designing small-scale, medium-scale, large-scale, and multisite indexer clustering.

This chapter covered a hefty portion of the Splunk Enterprise Certified Architect exam blueprint. You became familiar with

- **Module 2**: Project requirements—5%

- **Modules 3 and 4**: Infrastructure planning—10%

- **Module 16**: Multisite index cluster—5%

- **Module 18**: Search head cluster—5%

Multiple-Choice Questions

A. Which file configures a Splunk search peer using the configuration file?

1. server.conf

2. deployment.conf

3. output.conf

4. distsearch.conf

B. Which Splunk component performs indexing and responds to search requests from the search head?

1. Forwarder

2. Search peer

  3. License manager

  4. Search head cluster

C. Which of the following is a valid distributed search group?

  1. [distributedSearch:test] default=false servers=server1,server2

  2. [searchGroup:test] default=false servers=server1,server2

  3. [distributedSearch:test] default=false servers=server1:8089,server2:8089

  4. [searchGroup:test] default=false servers=server1:8089,server2:8089

D. Which Splunk component consolidates individual results and prepares reports in a distributed environment?

  1. Indexers

  2. Forwarder

  3. Search head

  4. Search peers

E. What hardware attribute needs to be changed to increase the number of simultaneous searches (ad hoc and scheduled) on a single search head?

  1. Disk

  2. CPUs

  3. Memory

  4. Network interface cards

F. Which Splunk component does a search head primarily communicate with?

1. Indexer

2. Forwarder

3. Cluster manager

4. Deployment server

Answers

A. 4

B. 2

C. 3

D. 3

E. 2

F. 1

References

- https://www.splunk.com/pdfs/technical-briefs/splunk-validated-architectures.pdf

- Splunk 7.x Quick Start Guide: Gain Business Data Insights from Operational Intelligence

- https://help.splunk.com/en/splunk-enterprise/administer/inherit-a-splunk-deployment/10.0/inherited-deployment-tasks/deployment-topologies

- `https://help.splunk.com/en/splunk-enterprise/administer/inherit-a-splunk-deployment/10.0/inherited-deployment-tasks/draw-a-diagram-of-your-deployment`

- `https://help.splunk.com/en/splunk-enterprise/administer/distributed-search/10.0/deploy-distributed-search/add-search-peers-to-the-search-head`

- `https://help.splunk.com/en/splunk-enterprise/administer/distributed-search/10.0/configure-search-head-clustering/configure-the-search-head-cluster`

- `https://help.splunk.com/en/splunk-enterprise/administer/distributed-search/10.0/deploy-search-head-clustering/deploy-a-search-head-cluster`

- `https://help.splunk.com/en/splunk-cloud-platform/splunk-validated-architectures/splunk-platform-indexing-and-search/topology-selection-guidance`

CHAPTER 14

Troubleshooting in Splunk

This chapter focuses on analyzing activity and diagnosing problems within Splunk Enterprise. You learn about the Monitoring Console used for troubleshooting and all its components. You get an overview of all the log files useful for troubleshooting. And you learn about the job inspector, what happens when a license violation occurs, and troubleshooting deployment and clustering issues.

This chapter covers the following topics:

- The Monitoring Console

- The log files for troubleshooting

- The metrics.log

- Job inspector

- Troubleshooting license violations

- Troubleshooting deployment issues

- Troubleshooting clustering issues

By the time you complete this chapter, you will have learned 20% of the Splunk architect exam blueprint.

© Carlos Moreno Buitrago, Deep Mehta 2026
C. M. Buitrago and D. Mehta, *The Splunk Core User Study Companion*, Certification Study Companion Series, https://doi.org/10.1007/979-8-8688-2501-9_14

Monitoring Console

The Monitoring Console in Splunk consists of a set of reports, dashboards, alerts, and health checks designed to provide a detailed view and information about your Splunk Enterprise performance. The Monitoring Console is available only to admin users and is one of the biggest troubleshooting assets. The following are a few of its most important functions:

- Single instance deployment

- Multi-instance deployment

- Monitors the system's health

- Configures forwarder monitoring

Single Instance Deployment Monitoring Console

The single instance deployment Monitoring Console is used for a single instance where the indexer, license manager, and search head are in one instance. To configure Splunk deployment for a single instance, refer to the following steps:

1. In Splunk Web, go to the Settings page.

2. Go to Monitoring Console.

3. Go to the Monitoring Console's Settings page, and then go to General Setup.

4. Click Standalone.

5. Select the indexer, license manager, KV Store, and Search Head under Server roles (see Figure 14-1).

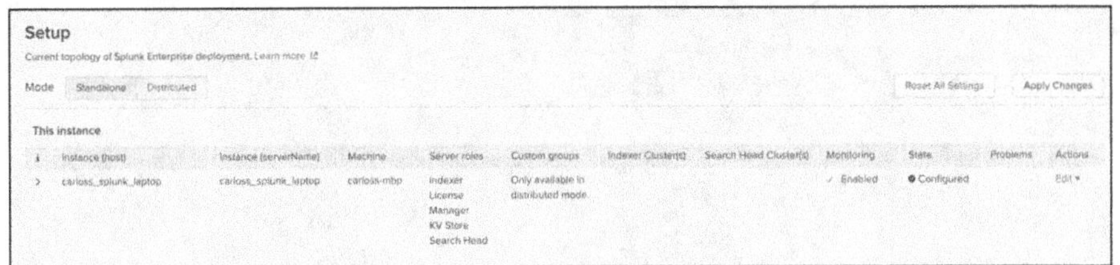

Figure 14-1. *Monitoring Console single instance*

Multi-instance Deployment Monitoring Console

The multi-instance deployment Monitoring Console is used for distributed environments where the indexer, license manager, and search head are not located within a single instance. The Monitoring Console will usually be in a separate Splunk instance. To configure Splunk deployment for a multi-instance, refer to the following steps:

1. In Splunk Web, go to the Settings page.

2. Go to the Monitoring Console.

3. Go to the Monitoring Console's Settings page, and then go to General Setup.

4. Turn on Distributed Mode.

5. Assign the roles to the node and configure the deployment.

6. Click Save.

Figure 14-2 shows how you edit the server roles.

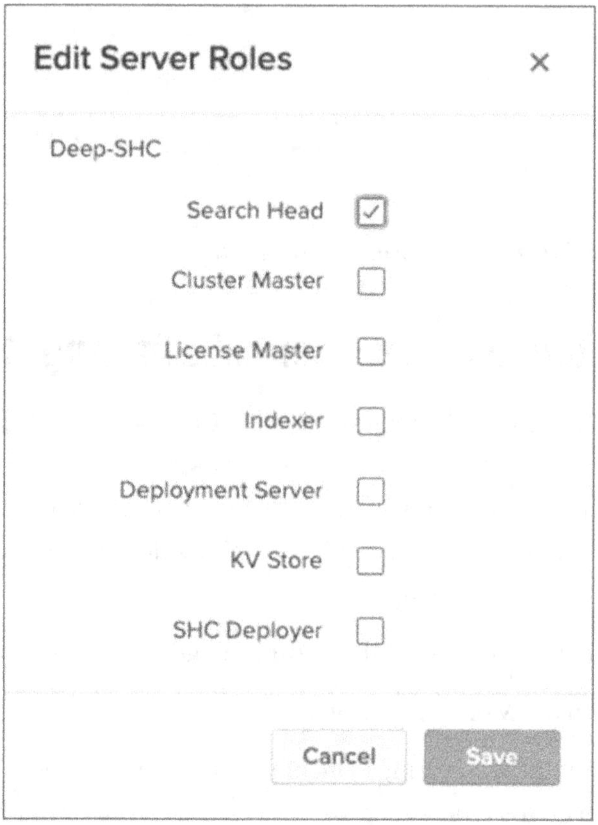

Figure 14-2. Monitoring Console server role

The following explains how to reset the server role after a restart:

1. In the Monitoring Console, click Settings and go to General Setup.

2. Click Distributed Mode.

3. To edit server roles, go to Remote Instance, and click Edit.

4. Select or edit the server role.

5. Click Apply changes.

Figure 14-3 shows how to edit the server role and assign roles to the instance.

Setup

Control topology of your Splunk Enterprise deployment. Learn more

Mode Standalone Distributed   Read Only

This Instance

i	Instance (host)	Instance (serverName)	Machine	Server roles	Custom groups	Indexer Cluster(s)	Search Head Cluster(s)	Monitoring	State
1	ds	ds	ip-172-31-79-228.ec2.internal	Deployment Server				✓ Enabled	● Configured

Remote Instances

9 Instances   Filter

8 Selected Instances ▾   25 Per Page ▾

i		Instance (host) *	Instance (serverName) *	Machine *	Server roles	Custom groups	Indexer Cluster(s)	Search Head Cluster(s)	Monitoring *	State *
>	☐	Deep-SHC	Deep-SHC	ip-172-31-79-228.ec2.internal	Search Head		S2XCSH2-7YES-4KB-W63-CG1822066C1	P5X23XP-95C7-4QH-8086-SCOAC2A183D7	✓ Enabled	● New
>	☐	cluster	cluster	ip-172-31-76-106.ec2.internal	Cluster Master		S2XCSH2-7YES-4KB-W63-CG1822066C1		✓ Enabled	● Configured
>	☐	idx1	idx1	ip-172-31-79-68.ec2.internal	Indexer		S2XCSH2-7YES-4KB-W63-CG1822066C1		✓ Enabled	● New
>	☐	idx2	idx2	ip-172-31-68-16.ec2.internal	Indexer		S2XCSH2-7YES-4KB-W63-CG1822066C1		✓ Enabled	● New
>	☐	idx3	idx3	ip-172-31-35-215.ec2.internal	Indexer		S2XCSH2-7YES-4KB-W63-CG1822066C1		✓ Enabled	● New
>	☐	idx4	idx4	ip-172-31-43-227.ec2.internal	Indexer		S2XCSH2-7YES-4KB-W63-CG1822066C1		✓ Enabled	● New
>	☐	sh1	sh1	ip-172-31-74-232.ec2.internal	Search Head		S2XCSH2-7YES-4KB-W63-CG1822066C1	P5X23XP-95C7-4QH-8086-SCOAC2A183D7	✓ Enabled	● New
>	☐	sh2	sh2	ip-172-31-66-166.ec2.internal	Search Head		S2XCSH2-7YES-4KB-W63-CG1822066C1	P5X23XP-95C7-4QH-8086-SCOAC2A183D7	✓ Enabled	● New
>	☐	sh3	sh3	ip-172-31-70-49.ec2.internal	Search Head		S2XCSH2-7YES-4KB-W63-CG1822066C1	P5X23XP-95C7-4QH-8086-SCOAC2A183D7	✓ Enabled	● New

Figure 14-3. *Monitoring Console Distributed mode*

6. Once you have made all changes in the server role, click Save and continue to the Overview page.

7. Click Go to Overview (see Figure 14-4).

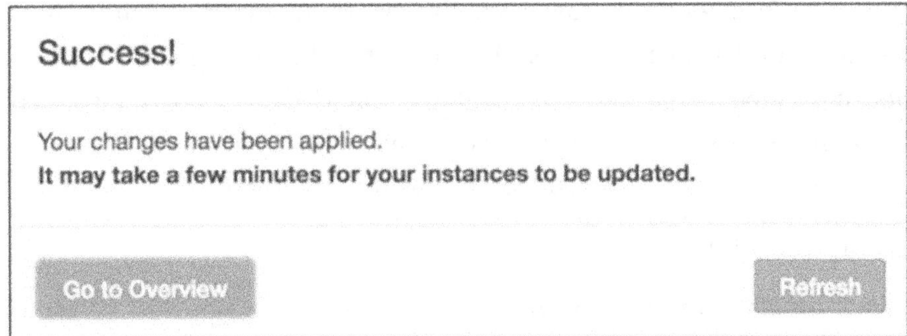

Figure 14-4. *Monitoring Console apply changes*

Figure 14-5 shows the multi-instance deployment Monitoring Console. There are four indexers, four search heads, and one cluster manager.

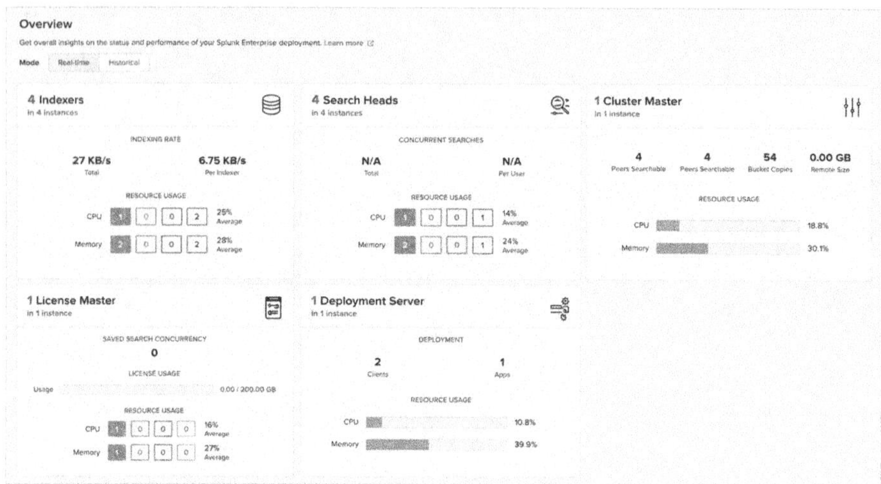

Figure 14-5. *Monitoring Console deployment overview*

Monitor System Health with Monitoring Console

The health check monitors overall system health. If a particular forwarder is decommissioned, it remains on the forwarder table until you remove it. The forwarder asset table removes decommissioned nodes from the system. Platform alerts are saved searches in Splunk that notify the Splunk administrator if any condition is triggered using the Monitoring Console. There are a few preconfigured alerts that come with the health check. You can modify an existing health check or create a new one.

Do the following to run a health check in the system:

1. In Splunk Web, go to the Settings page.

2. Go to Monitoring Console.

3. Go to Health Check.

4. Start the health check.

Figure 14-6 shows a health check in the system. If you encounter an error, you need to start troubleshooting it. The health check helps you understand the current status of your system.

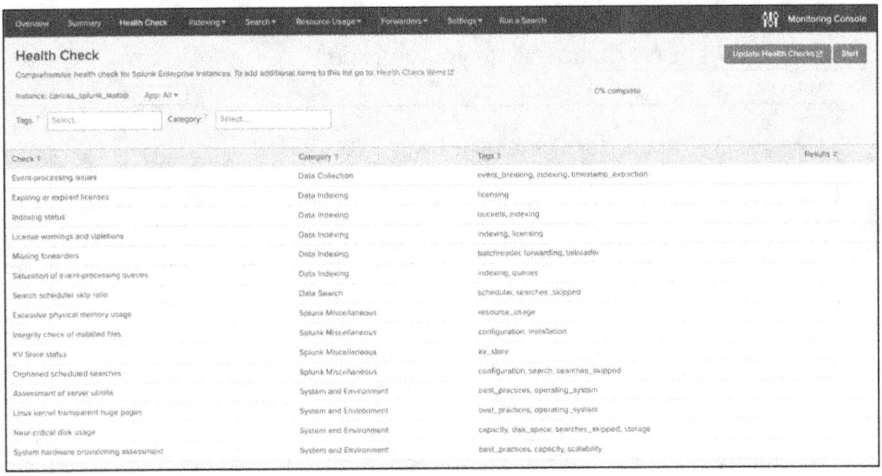

Figure 14-6. Monitoring Console Health Check

The forwarder monitoring dashboards summarize forwarder activity and throughput. When you enable forwarder monitoring, Splunk Enterprise turns on a scheduled search called "DMC Forwarder – Build Asset Table". That search uses the internal metrics written by the indexers. Be aware: in environments with large numbers of forwarders, this job can add noticeable load to the indexer search capacity. Do the following to rebuild a forwarder asset table:

1. In the Monitoring Console, click Settings.

2. Go to Forwarder Monitoring Setup.

3. Click Enable Forwarding if it's disabled.

4. Click Save.

5.   Click Rebuild Forwarder Asset Table.

6.   Select Timeline.

7.   Click Start Rebuild.

Figure 14-7 shows the message you will see.

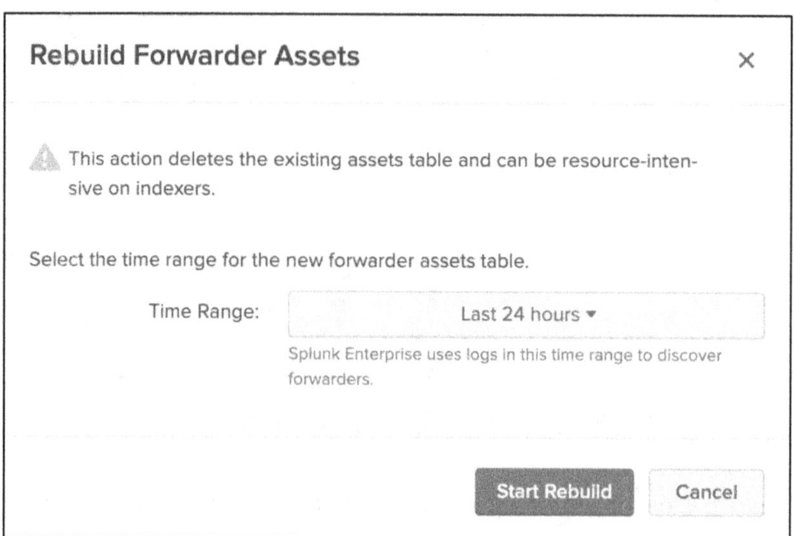

Figure 14-7. *Monitoring Console Rebuild Forwarder Asset Table*

The Monitoring Console removes missing forwarders by rebuilding the forwarder asset table.

Configure Forwarder Monitoring for the Monitoring Console

The Monitoring Console monitors incoming data from forwarder connections. The measurement of incoming data by a forwarder is done using the **metrics.log.**

The following explains how to set up a forwarder instance view:

1. In Splunk Web, go to the Settings page.

2. In the Monitoring Console, click Forwarders.

3. Go to Forwarder Instance.

4. Select the instance name and select a time range.

Figure 14-8 shows a forwarder instance where the instance is dell-pc and the time range is **5 minute window**.

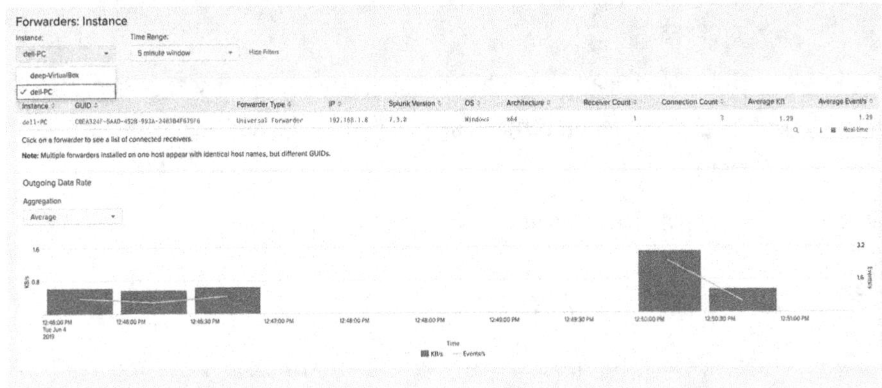

Figure 14-8. *Monitoring Console forwarder monitoring*

Next, let's look at log files for troubleshooting and metric analysis.

Log Files for Troubleshooting

The log files in Splunk are located at $SPLUNK_HOME/var/log/splunk/ and are used for troubleshooting Splunk Enterprise. The internal logs help you troubleshoot issues and are used for metric analysis. Table 14-1 covers all the major log files needed for troubleshooting. In this chapter, we will cover the metrics.log file in detail. Other log files can be found in `https://help.splunk.com/en/splunk-enterprise/administer/troubleshoot/10.0/splunk-enterprise-log-files/what-splunk-software-logs-about-itself.`

Table 14-1. *Internal log files for analysis and troubleshooting*

Attribute	Value
audit.log	This file consists of information about user activity and modifying knowledge objects and saved searches in Splunk.
btool.log	The Btool is used for troubleshooting Splunk configuration files. You use btool.log to log Btool activity.
metrics.log	This file consists of performance and system data, information about CPU usage by internal processors, and queue usage in Splunk's data processing.
splunkd.log	This file is used for troubleshooting. Any stderr messages generated by scripted input, scripted search command, and so on, are logged here.
splunkd_ access.log	This file is from splunkd through the logged UI.

Next, let's discuss the metrics.log file and the important group associated with it.

The metrics.log File

The metrics.log file reveals a picture of the entire Splunk Enterprise. Data in Splunk moves through the data pipelines in different phases. These phases include the input, parsing, merging, typing, and index pipelines, as shown in Figure 14-9.

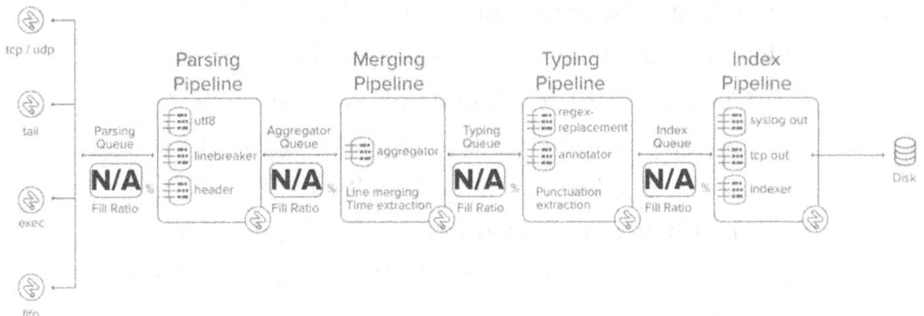

Figure 14-9. *Splunk Enterprise data pipeline flow*

- The **input pipeline** mainly consists of data sources. The host, source, and source type are added in this stage. This pipeline is the first phase. It includes the TCP/UDP, tailing, FIFO, and exec input methods. inputs.conf, wmi.conf, and props.conf are the configuration files in the input pipeline.

- The **parsing pipeline** is the second phase. This pipeline has two processors: **reader in** and **sendout**. The props.conf file is the file of interest in the parsing pipeline to transform the data stream and create individual events. The reader in the processor takes data from the parsing queue, and sendout sends data to the aggregator queue. Parsing UTF-8, line breakers, and headers are included in the parsing phase.

- The **merging pipeline** is the third phase. The event boundary detection and timestamp extraction occur in the merging stage, driven mainly by props.conf (with occasional transforms.conf). The aggregator processor determines where events start and end for multiline events. Some props configuration can be SHOULD_LINEMERGE, TIME_PREFIX, TIME_FORMAT, and MAX_TIMESTAMP_LOOKAHEAD to control multi-line handling and time extraction.

- The **typing pipeline** sits after parsing and before indexing and is where events get their final metadata (host, source, sourcetype, index), index-time actions run, and any allowed index-time field work occurs (e.g., INDEXED_EXTRACTIONS for CSV/JSON/TSV). This stage is driven mainly by props.conf and transforms. conf to route or rewrite and SEDCMD to edit raw events before they're stored.

- The **index pipeline** is the last phase. Events leave the typingQueue, enter the indexQueue, and are written to the target index's hot buckets (rawdata + tsidx) with their final metadata. Behavior here is driven mainly by indexes.conf (home/cold/frozen paths), volumes, retention rules, bucket sizing, and replication settings. For monitoring, rely on _internal metrics like index_ thruput to watch sustained write rates and latency.

When a data source suddenly surges, and it's hard to spot in the noise, start with metrics.log. Splunk writes this internal log about every 30 seconds, summarizing throughput, queues, and connection health. You can query it in Splunk Web with index=_internal source=*metrics.log or read it on disk at $SPLUNK_HOME/var/log/splunk/metrics.log. Events are organized by a group field that tells you what the line measures. The most useful groups for surge troubleshooting are

- **per_source_thruput**: Top talkers by source

- **thruput**: Overall ingest rates

- **index_thruput**: Write rates per index

- **queue**: Pipeline pressure/blocks

- **tcpin_connections**: Forwarder connection counts

For more information on metrics.log, refer to `https://help.splunk.com/en/splunk-enterprise/administer/troubleshoot/10.0/splunk-enterprise-log-files/troubleshoot-inputs-with-metrics.log`.

Pipeline Messages

Pipeline messages consist of detailed reports on the Splunk pipelines. They are placed together to process and manipulate events flowing in and out of the Splunk system. A pipeline message tracks the time used by each CPU and understands process execution.

The structure of a pipeline message is determined in the following event:

```
01/1/25
5:17:10.335 PM 01-01-2025 17:17:10.335 +0530 INFO Metrics -
group=pipeline, name=typing, processor=metricschema, cpu_
seconds=0, executes=405, cumulative_hits=8314
```

- **Timestamp**: The event consists of a timestamp that logs when the event occurred in the system.

- **group=pipeline**: This reflects and catches the CPU sendout execution time. Plotting total CPU seconds by processor shows you where the CPU time is going in indexing activity.

Queue Messages

Queue messages in Splunk consist of aggregating across events and can tell you where the indexing bottlenecks are located.

The structure of a queue message is determined in the following event:

```
01/1/25
8:25:19.839 PM
01/01/2025 20:25:19.839 +0530 INFO Metrics - group=queue,
blocked=true ,name=typingqueue, max_size_kb=500, current_size_
kb=432, current_size=955, largest_size=966, smallest_size=966
```

- **Timestamp**: The event consists of a timestamp that logs when the event occurred in the system.

- **group=queue**: Queue data consists of aggregates that reveal bottlenecks in the network. current_size is an important value as it reflects the amount of data inserted in the indexer at that time. For example: Currently, current_size=955, which reflects the amount of data inserted in the indexer is more at that time, which would lead to a bottleneck, and ideally, current_size should be between 0 and 500, which means the indexing system is working fine.

Thruput Messages

The thruput messages in Splunk measure the rate at which raw events are processed in the indexing pipeline. Throughput numbers are the size of the "raw" events traversing through the system.

The structure of a message is determined by the following event:

```
01/1/25
11:10:52.469 PM
01-01-2025 23:10:52.479 +0530 INFO Metrics - group=thruput,
name=thruput, instantaneous_kbps=3.3666908659618975,
```

```
instantaneous_eps=16.99733361182728, average_
kbps=2.13466754756423744, total_k_processed=46162,
kb=104.3837890625, ev=527, load_average=1.48828125
```

- **Timestamp**: The event consists of a timestamp that logs when an event occurred in the system.

- **group=thruput**: This measures the rate at which raw events are processed. It is best for insights during tuning performance or evaluating an average catchall indexing since Splunk is started. Thruput helps determine which data categories are busy and helps with performance tuning or evaluating indexing load.

Tcpout Connection Messages

Tcpout connection messages in Splunk provide information about the systems that the tcpout component is connected to via socket.

The structure of a tcpout connection message is determined in the following event:

```
01/1/25
11:10:52.469 PM
01-01-2025 23:10:52.479 +0530 INFO Metrics - group=tcpout_
connections, name=undiag_indexers:191.11.41.67:9997:0,
sourcePort=9009, destIp=191.11.41.67 destPort=9997, _tcp_
Bps=28339066.17, _tcp_KBps=21669.07, _tcp_avg_thruput=29532.96,
_tcp_Kprocessed=808751, _tcp_eps=31211.10, kb=807353.91
```

- **Timestamp**: The event consists of a timestamp that logs when the event occurred in the system.

- **group=tcpout_connections:** tcpout_connections measure the number of bytes that Splunk has successfully written to the socket. tcp_avg_thruput demonstrates the average rate of bytes flowing in the Splunk system since it started.

udpin_connections Messages

udpin_connections messages in Splunk are essentially metering the UDP input.

The structure of udpin_connections messages is determined by the following event:

```
01/1/25
11:10:52.469 PM
01-01-2025 23:10:52.479 +0530 INFO Metrics -group=udpin_
connections, 8001, sourcePort=8001, _udp_bps=0.00, _udp_
kbps=0.00, _udp_avg_thruput=0.00, _udp_kprocessed=0.00, _udp_
eps=0.00
```

- **Timestamp**: The event consists of a timestamp that logs when the event occurred in the system.

- **group=udpin_connections:** udpin_connection measures the bytes that Splunk has written to the UDP input. udp_avg_thruput demonstrates the average rate of bytes flowing in the UDP input since it started. It helps you understand the flow of data in your current system.

Next, let's look at Splunk search job inspectors to troubleshoot search performance.

Job Inspector

The Splunk search job inspector troubleshoots search performance and helps to understand the usage of the knowledge objects. The search job inspector provides a detailed summary report on Splunk's search performance.

How does a job inspector help you troubleshoot? The key information that is displayed by a search job inspector is execution cost and search job properties that play a major role in troubleshooting.

- **Execution cost** displays information about the components of the search and the amount of impact each component has on the search's output.

- **Search job** properties list the characteristics of the search. Each command in the search job has a part and a parameter. The value of the parameter reflects the time spent on each command.

- **EPS (Events Per Second)** shows the events per second so that you can understand the system's processing power and pipelines (EPS=total events/time).

Job Inspector Example Query

In this section, you run an example query to get group throughput events from the internal index and troubleshoot the query using a job inspector.

```
index="_internal" "group=thruput"
```

An execution cost query is shown in Figure 14-10.

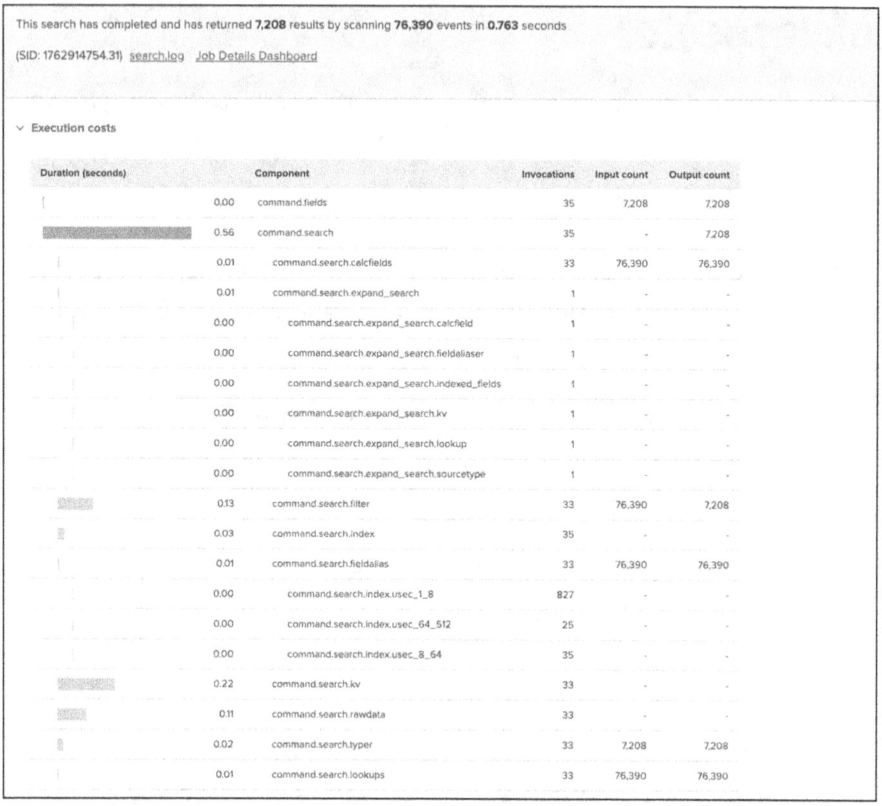

This search has completed and has returned **7,208** results by scanning **76,390** events in **0.763** seconds

(SID: 1762914754.31) search.log   Job Details Dashboard

Execution costs

Duration (seconds)	Component	Invocations	Input count	Output count
0.00	command.fields	35	7,208	7,208
0.56	command.search	35	-	7,208
0.01	command.search.calcfields	33	76,390	76,390
0.01	command.search.expand_search	1	-	-
0.00	command.search.expand_search.calcfield	1	-	-
0.00	command.search.expand_search.fieldaliaser	1	-	-
0.00	command.search.expand_search.indexed_fields	1	-	-
0.00	command.search.expand_search.kv	1	-	-
0.00	command.search.expand_search.lookup	1	-	-
0.00	command.search.expand_search.sourcetype	1	-	-
0.13	command.search.filter	33	76,390	7,208
0.03	command.search.index	35	-	-
0.01	command.search.fieldalias	33	76,390	76,390
0.00	command.search.index.usec_1_8	827	-	-
0.00	command.search.index.usec_64_512	25	-	-
0.00	command.search.index.usec_8_64	35	-	-
0.22	command.search.kv	33	-	-
0.11	command.search.rawdata	33	-	-
0.02	command.search.typer	33	7,208	7,208
0.01	command.search.lookups	33	76,390	76,390

Figure 14-10. *Splunk search job inspector execution cost*

- Duration includes the time needed to process each component.

- Component includes the name of the component.

- Invocations include the number of times the component was invoked.

- Input and Output are the events input and output by each component.

Events per second for index="internal"
"group=thruput" 76390/0.763=100118

The performance cost query is shown in
Figure 14-11.

This search has completed and has returned **7,208** results by scanning **76,390** events in **0.763** seconds

(SID: 1762914754.31) search log   Job Details Dashboard

> Execution costs

∨ Search job properties

canSummarize	None
createTime	2025-11-11T21:32:34.000-05:00
cursorTime	1969-12-31T19:00:00.000-05:00
custom	{ [+] }
defaultSaveTTL	604800
defaultTTL	600
delegate	None
diskUsage	2572288
dispatchState	DONE
doneProgress	1
dropCount	None
eai:acl	{ [+] }
earliestTime	2025-11-04T21:00:00.000-05:00
eventAvailableCount	7208
eventCount	7208
eventFieldCount	35
eventIsStreaming	true
eventIsTruncated	None
eventSearch	search (index="_internal" "group=thruput")
eventSorting	desc
indexEarliestTime	1762440641
indexLatestTime	1762914732
isBatchModeSearch	None
isDone	true

Figure 14-11. *Splunk search job inspector performance cost*

- Performance cost displays the status of the search result, provides the event count, and ensures that an event search is performed on all indexes.

In the next section, you will learn how the Splunk admin can use the _internal index to troubleshoot violations due to improper connections between the license manager and peer nodes.

Troubleshooting License Violations

License violation occurs when you exceed the indexing volume allowed for your license. The license manager keeps track of license usage, so if license usage exceeds license capacity, Splunk issues a license violation warning. If you get five warnings in a rolling 30-day period, you violate the license.

What happens when you violate your license?

During the license violation period, Splunk software continues indexing data, but **using search** is blocked, restriction is also imposed on scheduled searches and reports, and searching the **_internal** index is not blocked.

To get a detailed license usage report, use _internal index and licenseusage.log.

Violation Due to an Improper Connection Between License Manager and Peer Node

- A license manager defines the license stack and pool, adds licenses, and manages licensing through a central location. The peer nodes report to the manager node for licensing distribution.

- A **license peer** is a Splunk Enterprise node that reports to the manager for access to Enterprise licenses.

To better understand the link between a license manager and license peer and how to troubleshoot if syncing is absent, you need to refer to the following steps:

- Check the last time the Splunk manager contacted a license peer by going to Settings ➤ Licensing.

- To determine if the manager is down, go to the search command in a search head, and type a query similar to the following code block:

```
index="_internal" "license manager"
```

- If your manager is down, the message's structure looks like the following code block:

```
01/1/25
5:17:10.335 PM 01-01-2025 17:17:10.335 +0000 ERROR
LMTracker - failed to send rows, reason='Unable to
connect to license manager=https://191.168.1.20:8089
Error connecting: Connect Timeout
```

- If you find the message in the preceding code block, try to restart the license manager or configure a secondary manager node. For 72 hours, there is no issue with data to be indexed, but after 72 hours, Splunk deployment is severely affected.

- If you cannot find a message like the one in the preceding code block, restart your Splunk instance and check for a firewall policy; something in your network is wrong.

Next, let's look at troubleshooting deployment issues.

Troubleshooting Deployment Issues

Splunk deployment consists of forwarders that send data to indexers. Indexers index and save the data sent by forwarders. Search heads search data saved in a Splunk indexer. The forwarders, indexers, and search heads form a basic Splunk deployment. It's easy to troubleshoot a simple deployment with only a few forwarders, but to troubleshoot in an environment with thousands of forwarders could be complex if we do not know what to check. Let's look at troubleshooting different deployment layers.

Troubleshooting Splunk Forwarders

Splunk forwarders are instances that forward data to a Splunk indexer or heavy forwarder. To troubleshoot a Splunk forwarder, you need to check the internal logs using a search head, or if it does not send its logs due to connectivity issues with the indexers, go to the forwarder's splunkd.log file.

Where to Look First

- In Splunk Web:

  ```
  index=_internal host=<host_to_check>
  (source=*splunkd.log OR source=*metrics.log)
  (TcpOutputProc OR tcpout OR queue OR connection)
  ```

- On the forwarder host:

  - **splunkd.log**: $SPLUNK_HOME/var/log/splunk/
    splunkd.log

  - **metrics.log**: $SPLUNK_HOME/var/log/splunk/
    metrics.log

The following is the structure statement for an incomplete Splunk-to-Splunk (cooked) connection:

```
5:17:10.335 PM 01-01-2025 17:17:10.335 +0400 warn
TcpOutputProc - Cooked connection to ip=191.162.1.20:9997
timed out
```

- **Timestamp**: The event consists of a timestamp that logs when the event occurred in the system.

- **Severity/Component:** `warn TcpOutputProc` → the forwarder's **TCP output processor.**

- **Message: "Cooked connection … timed out"** = the forwarder couldn't complete the Splunk-to-Splunk (cooked) handshake/write to the target **ip:port. "Cooked"** here means the Splunk S2S protocol (metadata-aware), **not** raw syslog.

How Do You Troubleshoot the Connectivity?

- To troubleshoot syncing between forwarders and indexers, check the [tcpout] stanza in outputs.conf file in the forwarder.

- Is the indexer listening on the correct port? Are you using the same IP:Port?

- Linux forwarders use netstat, iptables, and Telnet commands to troubleshoot and check whether ports are open; if not, then enable a firewall policy that allows the port for outbound connection.

- Windows forwarders use Telnet commands to check whether ports are open; if not, then enable a firewall policy that allows the port for an outbound connection.

- Check the outputs.conf file for syntax. Restart the Splunk forwarder instance after editing the configuration file in the forwarder.

- If you use DNS for resolving the indexer IPs, make sure the DNS is working as expected. You can use tools such as dig or nslookup.

Troubleshooting Splunk Indexers

A Splunk indexer receives events (often on 9997 via Splunk-to-Splunk cooked traffic), parses, and saves them to index buckets (hot ➤ warm ➤ cold ➤ frozen). It's not just a repository directory; it's the engine that stores and makes data searchable. To troubleshoot a Splunk indexer, you need to check the internal logs using a search head or the indexer's splunkd.log file.

Where to Look First

- In Splunk Web:

  ```
  index=_internal (source=*splunkd.log OR
  source=*metrics.log)
  (TcpInputProc OR TcpInputFd OR index_thruput OR
  queue OR ERROR OR WARN)
  ```

- On the indexer:

  - **splunkd.log**: $SPLUNK_HOME/var/log/splunk/splunkd.log

  - **metrics.log**: $SPLUNK_HOME/var/log/splunk/metrics.log

The following is the structure of a statement for TcpInputFd:

```
5:19:20.355 PM 01-01-2025 17:17:10.335 +0400 ERROR TcpInputFd -
SSL Error for fd from HOST:localhost.localdomain, IP:127.0.0.1,
PORT:42134
```

- **Timestamp**: The event consists of a timestamp that logs when the event occurred in the system.

- **Component:** `TcpInputFd` = per-connection file descriptor on the **receiving** side (indexer).

- **Meaning:** An SSL/TLS handshake problem on the receiver.

- **Frequent causes:** Wrong/expired cert, CA mismatch, TLS version/cipher mismatch, `requireClientCert=true` without a client cert, hostname/CN mismatch, or clock skew.

How Do You Troubleshoot TcpInputFd?

- Confirm the indexer is listening on the expected port, for example, 9997.

- Match TLS settings with forwarders: CA chain, server cert, allowed TLS versions/ciphers, etc.

- Check time and DNS.

- Recheck logs after changes: index=_internal source=*splunkd.log (TcpInputFd OR TcpInputProc).

- Other issues can be generated due to backpressure on the queues. Check them as shown in this chapter or using the Monitoring Console.

- Always make sure the firewall configuration is working as expected.

Troubleshooting Clustering Issues

In a Splunk cluster, there are search heads, managers, and indexers. If a cluster is multisite, each cluster has a component. Although the manager node resides in any physical site, the manager node isn't part of any site.

- The **manager node** is responsible for managing the entire cluster. It coordinates the peer nodes' replicating activity and guides the search head in finding data in the peer node.

- The **peer node** (indexer) receives and indexes incoming data. It works as a standalone machine, but there is one difference: a peer node replicates data coming from another node. The peer node is responsible for storing internal data and copying received data from another node.

- The **search head** runs searches across the set of peer nodes. The manager node helps the search head find the peer nodes. Search heads are used to search across the entire Splunk environment.

Troubleshooting clustering is complex because there are a series of operations involved.

To troubleshoot clustering issues in Splunk, the admin generally uses a Monitoring Console to check for any issues.

How Do You Troubleshoot Cluster Issues?

- Check the cluster health: Check Replication/Search Factors met, peer states (Up/Down), fix-ups in progress, and searchable primaries.

- Make sure the peers receive CM heartbeats: heartbeat failures, version/bundle mismatches, or blocked replication.

- Double-check network and/or TLS issues.

- Check the bundle and bucket status: Make sure the search heads and indexers have enough resources (CPU/storage) to perform the maintenance cluster tasks.

Search Affinity

When you use multisite clustering, search heads prefer local-site peers to get search results and minimize cross-site traffic; this is called search affinity. If your local site loses peers, searches may struggle to meet search factor (SF), relying only on local primaries.

Emergency workaround (temporarily disable local bias): On the search head, set it to a site-less identity so it won't prefer the local site:

```
# server.conf on the SH
[general]
site = site0

[clustering]
multisite = true
manager_uri = https://<cm-host>:8089
mode = searchhead
```

Restart the SH. With site = site0, the SH will pull from any healthy site to meet SF. Use it with caution since it reintroduces cross-site latency and bandwidth usage. Use as a temporary measure until the local site stabilizes.

More Bucket Issues

In Splunk, data is grouped into buckets. Very large clusters (>40 peers and tens of thousands of buckets) put more load on the cluster control plane. If you see frequent heartbeat or control-message timeouts even after fixing storage/network bottlenecks, it's likely time for you to increase the heartbeat interval time for smoother performance. We recommend you consult the Splunk support team to determine whether it's safe for your organization. If advised, you can adjust on the CM:

```
# server.conf on the CM
[clustering]
heartbeat_timeout=120
```

Summary

This chapter covered Monitoring Console and all its important components, including single instance and multi-instance deployment, monitoring system health, and forwarder monitoring. It gave an overview of the metrics. log file, and you saw how a job inspector troubleshoots a Splunk search. At the end of the chapter, you saw various methods to troubleshoot license violations, deployment issues, and clustering issues in Splunk.

This chapter covered the following topics of the Splunk Enterprise Certified Architect exam blueprint:

- **Module 8**: Splunk Troubleshooting Methods and Tools—5%

- **Module 9**: Clarifying the Problem—5%

- **Module 12**: Search Problems—5%

- **Module 13**: Deployment problems—5%

The next chapter covers deploying apps through a deployment server, creating a server group using serverclass.conf, deploying an app on the universal forwarder through a deployment server, load balancing, the SOCKS proxy, and indexer discovery.

Multiple-Choice Questions

A. How does the Monitoring Console monitor forwarders?

1. By pulling internal logs from forwarders

2. By using the forwarder monitoring add-on

3. With internal logs forwarded by instances and API calls

4. With internal logs forwarded by the deployment server and API calls

B. How do you remove missing forwarders from the Monitoring Console?

1. By restarting Splunk

2. By rescanning active forwarders

3. By reloading the deployment server

4. By rebuilding the forwarder asset table

C. What is the default expiration time of a
    search job ID?

    1. 10 minutes

    2. 60 minutes

    3. 7 minutes

    4. 100 seconds

D. Which log files are useful in troubleshooting? (Select
    all options that apply.)

    1. _internal

    2. metrics.log

    3. _introspection

    4. splunkd.log

E. The Monitoring Console is useful in troubleshooting
    clustering issues.

    1. True

    2. False

F. Given a "warn TcpOutputProc - Cooked connection
    to ip:9997 timed out...", what is the most likely
    corrective action?

    1. Turn off useACK on the forwarders.

    2. Fix TLS trust: correct CA/chain and cert CN/SANs, align TLS
        settings, and ensure time sync.

    3. Disable the receiver on the indexer until the load decreases.

    4. Raise heartbeat_timeout on the cluster manager to mask
        control-plane delays.

G. Which command is correct to run on the cluster
    manager to get the RF/SF?

1. splunk show shcluster-status

2. splunk show cluster-status

3. splunk list peer-nodes

4. splunk display replication-state

Answers

A. 3

B. 4

C. 1

D. 2, 4

E. 1

F. 2

G. 2

References

- https://help.splunk.com/en/splunk-enterprise/
  search/spl-search-reference/10.0/search-
  commands/cluster

- https://help.splunk.com/en/splunk-enterprise/
  search/spl-search-reference/10.0/search-
  commands/cluster

- https://help.splunk.com/en/splunk-enterprise/
  get-started/overview/10.0/splunk-enterprise-
  resources-and-documentation/troubleshooting

- https://help.splunk.com/en/splunk-enterprise/
  search/spl-search-reference/10.0/search-
  commands/cluster

- https://help.splunk.com/en/splunk-enterprise/
  administer/troubleshoot/10.0/splunk-enterprise-
  log-files/what-splunk-software-logs-
  about-itself

- https://help.splunk.com/en/splunk-enterprise/
  administer/distributed-deployment-manual/10.0/
  overview-of-splunk-enterprise-distributed-
  deployments/how-data-moves-through-splunk-
  deployments-the-data-pipeline

- https://help.splunk.com/en/splunk-enterprise/
  administer/troubleshoot/10.0/splunk-enterprise-
  log-files/about-metrics.log

Advanced Deployment in Splunk

In this chapter, you will work with deployment servers to manage and distribute apps at scale. You will create deployment apps, organize clients into server classes, and push configurations to forwarders through agent management. You will also see how to deploy apps in clustered environments using the cluster manager and how to make search-related content available across your environment.

On the connectivity side, you will learn how load balancing works on forwarders (by time and by volume), how indexer discovery lets forwarders automatically find and use cluster peers, and how a SOCKS proxy fits into environments where direct network access to indexers is restricted.

In summary, this chapter covers the following topics:

- Deploying an app through a deployment server

- Creating a server group using serverclass.conf

- Deploying an app using a cluster manager

- Deploying a search app using a deployment server

- Load balancing

- Indexer discovery

- The SOCKS proxy

© Carlos Moreno Buitrago, Deep Mehta 2026

C. M. Buitrago and D. Mehta, *The Splunk Core User Study Companion*, Certification Study Companion Series, https://doi.org/10.1007/979-8-8688-2501-9_15

Deploying Apps Through the Deployment Server

In this section, you will learn how to add apps to the deployment server (DS) to be configured on the forwarders.

A deployment app in Splunk consists of arbitrary content that you want to install in the deployment clients. Arbitrary content can be a Splunk Enterprise app, a set of Splunk configurations, and scripts. You can create a directory for the app. After creating the directory, you can add scripts, configuration files, images, and so forth. Anything you'd package as a Splunk app:

- `default/` and `local/` with *.conf (e.g., inputs.conf, props.conf, transforms.conf)

- optional `bin/` scripts, `metadata/`, lookups, dashboards

Create App Directories and View Apps in Agent Management

Creating app directories and viewing apps in the agent management are the most important tasks in Splunk.

Create App Directories

To create directories in the deployment server for each deployment app, go to the `$SPLUNK_HOME/etc/deployment-apps` folder. There, create a directory for your app.

In this case, the certification_book_forwarders app was created at `$SPLUNK_HOME/etc/deployment-apps`. The name of the directory serves as the name of the app within agent management. You can add apps in

Splunk agent management at any time. After creating an app, you can view it on Splunk Web or run a CLI command. The following CLI command reloads the deployment server configuration:

```
splunk reload deploy-server
```

View Apps in Agent Management

Agent management plays the most important role in Splunk deployment. It provides an interface to create server classes, which are a group of Splunk apps assigned to forwarders. You can create as many server classes as you need, and they map forwarders to apps. To view apps in the agent management, refer to the following steps:

1. Go to the Settings page from Splunk Web, and go to Agent Management.

2. In Applications, you can find the app name. You should see the app named certification_book_ forwarders. (If not, then in Splunk CLI, type the following command; also refer to Figure 15-1.)

    ```
    splunk reload deploy-server
    ```

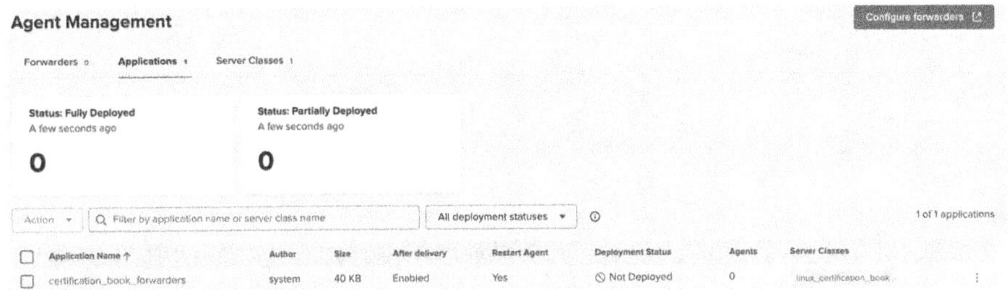

Figure 15-1. *Apps in the Agent Management*

Creating a Server Group Using serverclass.conf

Agent management in Splunk provides an interface to create server classes and map forwarders to deployment apps. The interface saves server configurations to the serverclass.conf. In some advanced server class configurations, you might need to directly edit serverclass.conf.

Configure a Server Class Using Agent Management

Server classes map forwarders to deployment apps. To create a server class, refer to the following steps:

1. Go to Settings and go to Agent Management.

2. Go to Server Classes, and click the New Server Class button.

3. In Server Class, set Name=linux_certification_book (see Figure 15-2).

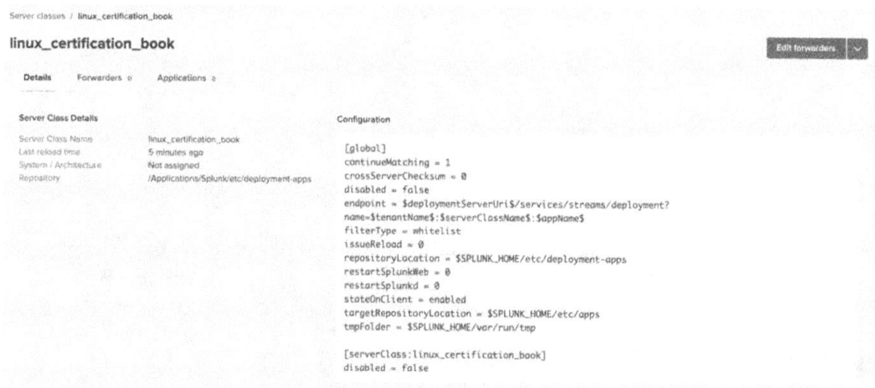

Figure 15-2. *linux_certification_book server class*

4. Go to Applications, and click Edit applications.

    Select the certification_book_forwarders as an assigned application as shown in Figure 15-3.

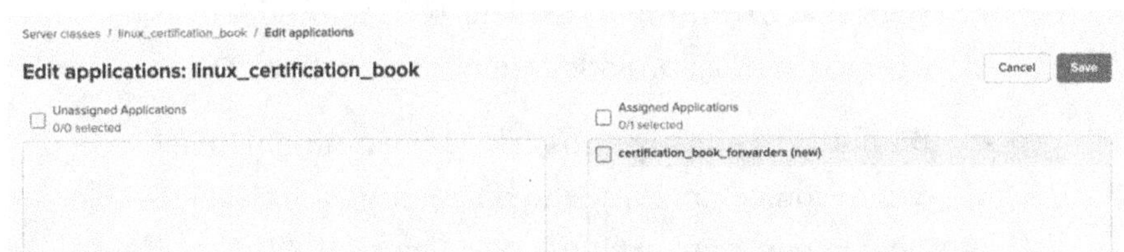

Figure 15-3. *certification_book_forwarders app assigned to a server class*

5. Select the certification_book_forwarders app, and make sure the app and the Restart agent options are enabled, as shown in Figure 15-4.

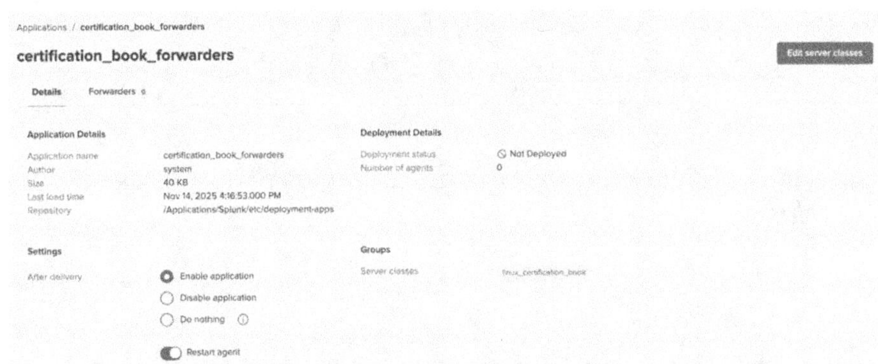

Figure 15-4. *certification_book_forwarders app configurations*

6. Go to the linux_certification_book server class and to the Forwarders page.

7. Click Edit forwarders.

8. In the forwarders, you find a whitelist and blacklist block, followed by all clients that are polling the deployment server.

    - **Whitelist in adding apps**: This is a required field in which you must provide the hostname, IP, or DNS.

    - **Blacklist in adding apps**: This is an optional field where you can provide the hostname, IP, or DNS of what you want to exclude. (See Figure 15-5.)

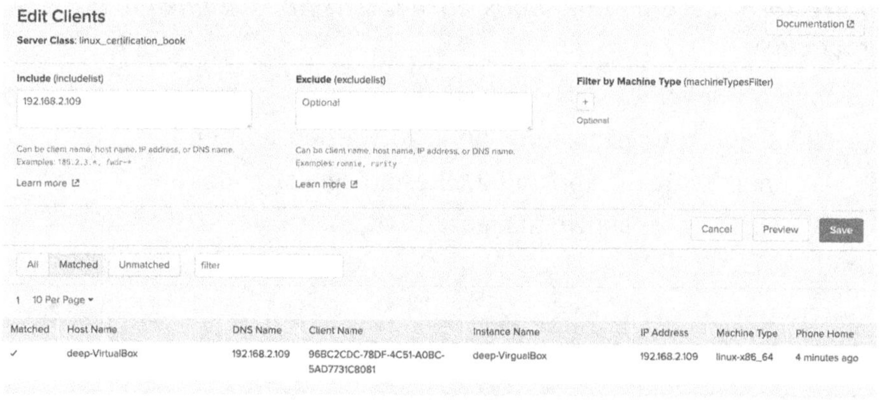

Figure 15-5. Forwarders in the certification_book_forwarders server class

9. Click Save.

Now, all the forwarders included in the whitelist will get a copy of the certification_book_forwarders app. If in the future you add more apps to the linux_certification_book server class, the forwarders will get them automatically. As mentioned, the deployment server will map the forwarders with the assigned apps, defined by the server classes.

Redeploy Apps to the Client

The deployment server is a centralized configuration manager that is responsible for deploying apps on client instances. Redeployment is generally done when you change the server class (changes in the set of apps or clients) or change an app's configuration file. Generally, redeployment is done automatically, but in some cases, you need to redeploy using serverclass.conf or an app using agent management.

Deploy an App by Editing serverclass.conf

You can also create server classes by editing the serverclass.conf file. The deployment server does not automatically reload the configuration, so the changes won't be applied to the client. To reload the configuration and deploy a new app in the deployment server to the forwarders, you need to reload the configuration by running the following CLI command:

```
splunk reload deploy-server
```

Redeploy an App After You Change the Content

When you add a script or a configuration file to your existing app in Splunk, it's necessary to reflect the changes in all your client machines. To redeploy an app with updated content, you need to follow these instructions:

1. Update the content in the deployment app directory on the deployment server.

2. Reload the configuration in the deployment server with the command shown before.

Deploy Apps to New Forwarder

When a new forwarder is registered with the deployment server, it will automatically pull every app that its server classes assign to it. This happens on the client's next phone-home cycle (controlled by phoneHomeIntervalInSecs). Just make sure the client can reach the DS on 8089 port (or a custom one) and that it matches the server class whitelist/blacklist rules.

App Management Issues

The deployment server is very useful for managing apps, but there are some issues linked to app management. Before you deploy an app, let's discuss issues linked to app management in Splunk.

Deployment Server Is Irreversible

Once an app is managed by the deployment server, the DS is authoritative for that app on its deployment clients. If you unassign or remove the app from the DS (or its server class), the client will uninstall it on the next phone-home. That action does not tell the client to self-manage the same app; it only tells it to remove the DS-managed copy. If you want to stop DS control:

- **For a specific app, but keep it installed**: Remove it from DS/server class, then install a local, unmanaged copy on the client (often with a different app name to avoid DS cleanup).

- **For the whole client**: Disable DS polling (remove/modify deploymentclient.conf or splunk set deploy-poll <none>) so it no longer calls the DS.

Apps with Lookup Tables

Some apps include scripts that write data into lookup tables. When those apps are managed by the deployment server, any app update pushed from the DS can overwrite the app's local content, including lookup files that the scripts have been updating. As a result, the forwarder may lose its locally generated data every time the forwarder pulls the correct version of the app from the DS.

To control this behavior, Splunk provides specific options in serverclass.conf such as excludeFromUpdate=app_root/lookups, and blacklist/whitelist controls for files and directories. These settings allow you to preserve selected local files, skip overwriting lookups, or prevent the DS from replacing certain parts of an app while still managing the rest.

Deploy Configuration File Through Cluster Manager

The cluster manager node in Splunk coordinates the peer node activities that regulate the functioning of the indexer cluster. The manager node controls and manages index replication, distributes app bundles to peer nodes, and helps search heads locate bucket copies on peers. (You learned how to configure the manager node in Chapter 9.)

Managing Indexes on the Indexer Using the Manager Node

To manage the indexes of all peer nodes, you need to use the same configuration file for all indexers. You do not need to go to each of the Splunk instances and edit its indexes file (it is not recommended to manually edit each indexer's configuration file). You need to use the manager node for the deployment of the same configuration file for all indexers. In this section, you deploy the same indexes.conf file on four indexers with two sites—site1 and site2—using the cluster manager.

Assuming the cluster was already configured on the cluster manager, go to the manager node (using Splunk Web), go to settings, and move to Indexer Clustering. There, you can see all the indexes that have been reported to the manager node. In this example, four indexers reported to the manager node. Figure 15-6 shows the instances that pinged the manager node.

Indexer Clustering: Master Node

	All Data is Searchable		Search Factor is Met		Replication Factor is Met
	4 searchable 0 not searchable		4 searchable 0 not searchable		2 searchable 0 not searchable
	Peers		Indexes		Indexes

Peers (4)   Indexes (4)   Search Heads (0)

Peer Name	Site	Fully Searchable	State	Buckets
idx3	site2	Yes	Up	8
idx1	site2	Yes	Up	8
idx2	site1	Yes	Up	8
idx4	site1	Yes	Up	8

Figure 15-6. *Manager node indexer clustering*

We will deploy an index called test across the cluster. You could configure it manually, but it is not a good practice. In this section, you use a cluster manager to deploy the indexes.conf file, where you enable a replication factor and configure the index.

1. Go to the cluster manager node. Traverse to
    `$SPLUNK_HOME/etc/manager-apps/_cluster/local`,
    and create indexes.conf file.

2. Edit indexes.conf according to the following code:

```
[test]
homePath=$SPLUNK_DB/test/db
coldPath=$SPLUNK_DB/test/colddb
thawedPath=$SPLUNK_DB/test/thaweddb
repFactor=auto
```

3. To deploy indexes.conf file, move to $SPLUNK_HOME/
    bin/ in the cluster manager.

```
./splunk validate cluster-bundle --check-restart
```

4. To confirm the status of bundle validation, refer to
    the following command:

```
./splunk show cluster-bundle-status
```

If your cluster bundle is valid, you find the checksum of your new bundle. The output will be similar to Figure 15-7.

```
[ec2-user@ip-172-31-75-109 bin]$ sudo ./splunk validate cluster-bundle  --check restart
Validating new bundle. Please run 'splunk show cluster-bundle-status' to check the status of the bundle validation.
[ec2-user@ip-172-31-75-109 bin]$ sudo ./splunk show cluster-bundle-status

master
        cluster_status=None
        active_bundle
                checksum=826AF3010CA7165419661216C27A2AAD
                timestamp=1592876799 (in localtime=Tue Jun 23 01:46:39 2020)
        latest_bundle
                checksum=826AF3010CA7165419661216C27A2AAD
                timestamp=1592876799 (in localtime=Tue Jun 23 01:46:39 2020)
        last_validated_bundle
                checksum=826AF3010CA7165419661216C27A2AAD
                last_validation_succeeded=1
                timestamp=1592918946 (in localtime=Tue Jun 23 13:29:06 2020)
        last_check_restart_bundle
                last_check_restart_result=restart not required
                checksum=
                timestamp=0 (in localtime=Thu Jan  1 00:00:00 1970)

  idx3    0700F4C5-9908-4036-AE0C-58A347989BAA     site2
          active_bundle=826AF3010CA7165419661216C27A2AAD
          latest_bundle=826AF3010CA7165419661216C27A2AAD
          last_validated_bundle=826AF3010CA7165419661216C27A2AAD
          last_bundle_validation_status=success
          restart_required_apply_bundle=0
          status=Up

  idx4    DC29F3C9-32C4-454C-8D53-8C0CAA618549     site2
          active_bundle=826AF3010CA7165419661216C27A2AAD
          latest_bundle=826AF3010CA7165419661216C27A2AAD
          last_validated_bundle=826AF3010CA7165419661216C27A2AAD
          last_bundle_validation_status=success
          restart_required_apply_bundle=0
          status=Up

  idx1    DCD95AC2-E49C-4CEB-A9D4-C54B00C2B832     site1
          active_bundle=826AF3010CA7165419661216C27A2AAD
          latest_bundle=826AF3010CA7165419661216C27A2AAD
          last_validated_bundle=826AF3010CA7165419661216C27A2AAD
          last_bundle_validation_status=success
          restart_required_apply_bundle=0
          status=Up

  idx2    FEB0AD2C-4FEF-43DC-83D1-3230BFCD80EE     site1
          active_bundle=826AF3010CA7165419661216C27A2AAD
          latest_bundle=826AF3010CA7165419661216C27A2AAD
          last_validated_bundle=826AF3010CA7165419661216C27A2AAD
          last_bundle_validation_status=success
          restart_required_apply_bundle=0
          status=Up
```

Figure 15-7. *Validating cluster bundle*

5. Once the bundle was validated, you can apply it
    to the peers, so all the indexers will get the same
    copy of the configuration files. Run the following
    command:

```
./splunk apply cluster-bundle
```

To monitor the cluster manager status, navigate to Splunk Web settings, and go to Indexer Clustering, where you find the index test's status (see Figure 15-8).

Indexer Clustering: Master Node

✓ All Data is Searchable   ✓ Search Factor is Met   ✓ Replication Factor is Met

4 searchable   **0** not searchable
Peers

4 searchable   **0** not searchable
Indexes

4 searchable   **0** not searchable
Indexes

Peers (4)   **Indexes (4)**   Search Heads (0)

Bucket Status ▾   Filter   🔍   10 per page ▾

Index Name	Fully Searchable	Searchable Data Copies	Redundant Data Copies	Buckets	Earliest Live Data Max
_audit	✓ Yes	2	2	6	~1.28 d ago
_internal	✓ Yes	2	2	8	~1.28 d ago
_introspection	✓ Yes	2	2	8	~1.28 d ago
test	✓ Yes	2	2	8	~1.28 d ago

Figure 15-8. *Indexer clustering status*

Deploy App on Search Head Clustering

The deployer distributes apps to the search head cluster. The deployer is a Splunk instance that distributes apps and configuration updates to each search head cluster member. The set of updates that the deployer distributes is called configuration bundles. The main task of the deployer is to handle the apps and user configuration which the SHC members pull and apply in a coordinated way. The deployer's primary role is to deliver baseline, non-runtime, non-replicated configurations, things like apps, default settings, lookups, UI assets, and knowledge objects that are not created or modified at search time.

The deployer does not run searches, hold captaincy, or participate in replication; it simply ensures that every search head in the cluster starts from the same, clean, synchronized app baseline.

Configure the Deployer to Distribute Search Head Apps

In this section, you use a deployer to push the SHC (search head cluster) app to search head clustering nodes. You already saw how to configure search head clustering. In this section, you learn how to deploy the solution.

1. In the deployer, create an app named shc located in $SPLUNK_HOME /etc/shcluster/apps/.

2. Create or edit app.conf located in /local/app.conf of the shc app. Refer to the following code:

    ```
    [ui]
    is_visible = 0
    [package]
    id = shc_base
    check_for_updates = 0
    ```

3. To push the configuration to the search head clustering nodes, move to $SPLUNK_HOME/bin/ on the deployment server and refer to the following command (see Figure 15-9):

    ```
    splunk apply shcluster-bundle -action stage
    --answer-yes
    ```

```
[root@ip-172-31-77-218 bin]# sudo ./splunk apply shcluster-bundle -action stage --answer-yes
Your session is invalid.  Please login.
Splunk username: admin
Password:
Bundle has been pushed successfully to all the cluster members.
[root@ip-172-31-77-218 bin]#
```

Figure 15-9. Deploying cluster bundle on search head clustering

4. The deployment is ready to be pushed. To do this, you need to contact the search head cluster captain and use the apply shcluster-bundle command. Refer to the following command (see Figure 15-10):

```
splunk apply shcluster-bundle -action send
-target <url>:<mgmt_port>/<host_name>:<mgmt_
port>  --answer-yes
```

```
[root@ip-172-31-77-218 bin]# sudo ./splunk apply shcluster-bundle -action stage --answer-yes
Your session is invalid.  Please login.
Splunk username: admin
Password:
Bundle has been pushed successfully to all the cluster members.
[root@ip-172-31-77-218 bin]#  sudo ./splunk apply shcluster-bundle -action send -target https://172.31.72.220:8089 --answer-yes
Bundle has been pushed successfully to all the cluster members.
[root@ip-172-31-77-218 bin]#
```

Figure 15-10. *Deploying cluster bundle on the search head using the search head cluster captain*

5. When you traverse to the search head clustering nodes, you can find shc app deployed on it. Refer to Figure 15-11. Similarly, you can find the app deployed.

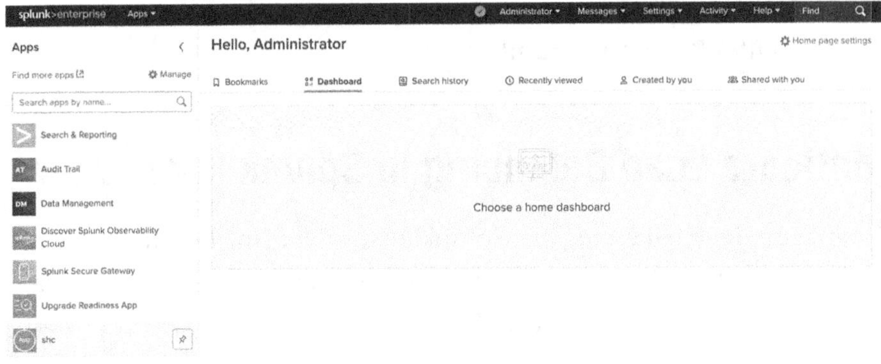

Figure 15-11. *shc app*

Load Balancing

The forwarders can distribute data across several Splunk indexers if needed to avoid overloading a single receiver, so each receiver receives a portion of the data stream. The forwarder routes data to the total available indexers at a specified time or volume interval. If you have three available indexers and a group of forwarders, the forwarders use a round-robin algorithm to switch the data stream to another index based on time or volume:

- **By time:** This factor controls how frequently forwarders switch from one indexer to another based on time. To enable the load balancer by time, update the settings in the universal forwarder's outputs.conf file by adding autoLBFrequency=<number>. autoLBFrequency is the most commonly used and recommended approach, and the default is 30 seconds.

- **By volume:** This factor controls how frequently forwarders switch from one indexer to another based on volume. To enable it, update the settings in the universal forwarder's outputs.conf file by adding autoLBVolume=<number>.

Configure Load Balancing in Splunk Forwarder

To configure load balancing in a Splunk forwarder using outputs.conf, you need to go to $SPLUNK_HOME/etc/system/local and create/edit outputs. conf or deploy a custom app with the outputs.conf file.

Time-Based Rotation (Recommended)

Go to $SPLUNK_HOME/etc/system/local/outputs.conf or deploy a
custom app on the forwarder, and edit the stanza according to the
following configuration, which sets autoLBFrequency by time. This factor
controls how frequently forwarders switch from one indexer to another.
Refer to the syntax for configuring the load balancing feature by time using
the following configuration file:

```
[tcpout:my_LB_indexers]
server = idx1.example.com:9997, idx2.example.com:9997
autoLB = true
autoLBFrequency = <seconds>
```

- server is the address of the indexer.

Volume-Based Rotation (Optional)

Go to $SPLUNK_HOME/etc/system/local/outputs.conf or deploy a
custom app on the forwarder, and edit the stanza according to the
following configuration, which sets autoLBVolume by volume. This factor
controls how frequently forwarders switch from one indexer to another.
Refer to the syntax for configuring the load balancing feature by volume
using the following configuration file:

```
[tcpout:my_LB_indexers]
server = idx1.example.com:9997, idx2.example.com:9997
autoLB = true
autoLBFrequency = <seconds>
```

Specify Load Balancing from Splunk CLI

To configure load balancing from Splunk CLI, you need to go to $SPLUNK_
HOME/bin on the forwarders and specify the commands. It is the easiest
way to load balance.

To specify load balancing from Splunk CLI, refer to the following:

```
splunk add forward-server idx1.example.com:9997 -method
autobalance
splunk restart
```

Indexer Discovery

Indexer discovery simplifies how forwarders connect to an indexer cluster
by letting the cluster manager (CM) tell them which peer nodes are
currently available. Each peer node reports its receiving port and status
to the CM. Forwarders are configured to poll the CM at regular intervals,
and every time they do, the CM returns a fresh list of healthy peers based
on its load-balancing policy. The forwarder then builds a tcpout group
automatically from that list and sends data to the peers without needing
hard-coded server entries in outputs.conf. This allows forwarders to adapt
automatically to rolling upgrades, peer failures, and scaling events.

Configure Indexer Discovery

To configure indexer discovery successfully, you need to configure the
components. In indexer discovery, peer nodes, the manager node,
and forwarders are the main components. You need to configure each
component as follows:

1. Configure the peer nodes to receive data from
    forwarders.

2. Configure the manager node to enable indexer
    discovery.

3. Configure the forwarders.

Configure the Peer Nodes

You can configure it either by editing the configuration files or by using
Splunk Web. Refer to the following instructions.

Configure Peer Node Using Splunk Web

To configure peer nodes to receive data from forwarders using Splunk Web,
follow these steps:

1. Go to Splunk Web, go to Settings, and move to
    Forwarding and Receiving.

2. In Configure Receiving, click Add New.

3. In Port, type **port address<port>**.

4. Click Save.

Configure the Peer Node Using inputs.conf

To configure peer nodes to receive data from forwarders using a
configuration file, go to `$SPLUNK_HOME/etc/system/local/inputs.conf`
and edit splunktcp stanza. Refer to the following splunktcp stanza:

```
[splunktcp://<port>]
disabled = 0
```

Configure the Manager Node

To configure the manager node to enable index discovery using a configuration file, go to $SPLUNK_HOME/etc/system/local/server.conf, and edit the indexer_discovery stanza. Refer to the following:

```
[indexer_discovery]
pass4SymmKey = <string>
polling_rate = <integer>
indexerWeightByDiskCapacity = <bool>
```

- **pass4SymmKey** in the manager node secures the connection between forwarders and the manager node.

- **polling_rate** adjusts the rate at which the forwarders poll the manager node.

- **indexerWeightByDiskCapacity** is an optional attribute that determines whether indexer discovery uses load balancing.

Configure the Forwarders

To configure the forwarders using a configuration file, go to $SPLUNK_HOME/etc/system/local/outputs.conf, and edit the indexer_discovery:<name> stanza. Refer to the following:

```
[indexer_discovery:<name>]
pass4SymmKey = <string>
manager_uri = <uri>

[tcpout:<target_group>]
indexerDiscovery = <name>

[tcpout]
defaultGroup = <target_group>
```

- **indexer_discovery:<name>** in the forwarder stanza sets <name> in the indexer discovery attribute.

- **pass4SymmKey** secures the connection between forwarders and manager nodes.

- **manager_uri** contains the address of the manager node.

- **tcpout:<target_group>** sets the indexer discovery attribute instead of the server attribute you use to specify the receiving peer nodes if you are not enabling indexer discovery.

SOCKS Proxy

The forwarder directly communicates with the indexer and sends data, but if the firewall configuration only allows traffic through a proxy, the forwarder cannot communicate with the indexer. In this scenario, you can configure a forwarder to use a SOCKS5 proxy host to send data to an indexer by specifying attributes in a stanza in the forwarder's outputs. conf file. The proxy host establishes a connection between the indexer and forwarder; then, the forwarder sends data to the indexer through the proxy host.

Configure SOCKS Proxy

A SOCKS proxy is configured by creating a [tcpout] or [tcpout-server] stanza within the outputs.conf file in the forwarder's $SPLUNK_HOME/ etc/system/local/outputs.conf file. Restart the forwarder after the configuration is updated.

Table 15-1 illustrates the configuration of a SOCKS proxy in the forwarder's outputs.conf file.

Table 15-1. *Configure SOCKS proxy using forwarder's outputs.conf*

Attribute	Value
socksServer=<ip/host>:<port>	Specifies the address of socks5 proxy where forwarder should connect for forwarding data.
socksUsername=<string>	Specifies the username to authenticate socks5 proxy. It is optional.
socksPassword=<string>	Specifies the password to authenticate socks5 proxy. It is optional.
socksResolveDNS=true|false	Specifies whether forwarder should use DNS to resolve the hostname of indexer in the output group.

The following code block illustrates an example of how a SOCKS proxy is configured in Splunk:

```
[tcpout]
defaultGroup = proxy_indexer_test

[tcpout:proxy_indexer_test]
server = <hostname1>:<port>,<hostname2>:<port>,......,<hostname n><port n>
socksServer = <proxy_name>:<port>
```

- **server** is the name of the server.

- **socksServer** specifies the address of socks5 proxy where the forwarder should connect to forward data.

Summary

In this chapter, you started the journey of deploying an app through a deployment server. You created app directories, deployed an app to the client through the deployment server, and pushed it to the client through forwarder management. You created a server group using serverclass. conf and pushed applications on a client using the server class in agent management. You also deployed an app on universal forwarder through the deployment server. In the next section, you learned how load balancing is implemented on Splunk based on time and volume. In a SOCKS proxy, you learned that if a firewall does not allow data to pass through the network, you can implement a SOCKS proxy. You also learned how indexer discovery is implemented to streamline the process of connecting forwarders to peer nodes in indexer clusters.

In this chapter, you covered a hefty portion of the Splunk Enterprise Certified Architect exam blueprint:

- **Module 6**: Forwarders and Deployment Best Practices—5%

- **Module 14**: Large-scale Splunk Deployment Overview—5%

- **Module 19**: Search Head Cluster Management and Administration—5%

In the next chapter, you will learn about managing indexes, managing index storage, managing index clusters, managing a multisite index cluster, Splunk REST API endpoints, and Splunk SDK.

Multiple-Choice Questions

A. Where should apps be located on the deployment server that the clients pull from?

1. $SPLUNK_HOME/etc/apps

2. $SPLUNK_HOME/etc/search

3. $SPLUNK_HOME/etc/manager-apps

4. $SPLUNK_HOME/etc/deployment-apps

B.   Which Splunk component distributes apps and certain other configuration updates to search head cluster members?

1. Deployer

2. Cluster manager

3. Deployment server

4. Search head cluster

C.   When running the following command, what is the default path in which the deploymentclient.conf is created?

```
splunk set deploy-poll deploy server:port
```

1. SPLUNK_HOME/etc/deployment

2. SPLUNK_HOME/etc/system/local

3. SPLUNK_HOME/etc/system/default

4. SPLUNK_HOME/etc/apps/deployment

D.   Which methods are used in load balancing? (Select all methods that apply.)

1. By time

2. By frequency

3. By volume

4. All of the above

E.   What are the main components of index discovery in Splunk? (Select all methods that apply.)

1. Peer node

2. Manager node

3. Universal forwarder

4. Search head

5. All of the above

6. None of the above

Answers

A. 4

B. 1

C. 2

D. 1, 3

E. 1, 2, 3

References

- ```
  https://help.splunk.com/en/splunk-enterprise/
  administer/update-your-deployment/10.0/deploy-
  apps/deploy-apps-to-agents
  ```

- ```
  https://help.splunk.com/en/splunk-enterprise/
  administer/update-your-deployment/10.0/deploy-
  apps/protect-content-during-app-updates
  ```

- ```
  https://help.splunk.com/en/splunk-enterprise/
  administer/update-your-deployment/10.0/
  configure-the-agent-management-system/create-
  server-classes
  ```
  ```

- https://help.splunk.com/en/splunk-enterprise/
  administer/manage-indexers-and-indexer-
  clusters/10.0/configure-the-peers/update-
  common-peer-configurations-and-apps

- https://help.splunk.com/en/splunk-enterprise/
  administer/distributed-search/10.0/update-
  search-head-cluster-members/use-the-deployer-
  to-distribute-apps-and-configuration-updates

- https://help.splunk.com/en/splunk-enterprise/
  administer/manage-indexers-and-indexer-
  clusters/10.0/get-data-into-the-indexer-
  cluster/use-indexer-discovery-to-connect-
  forwarders-to-peer-nodes

CHAPTER 16

Advanced Splunk

In this chapter, you learn about managing indexes, configuring custom indexes, removing indexes, and configuring index parallelization. You see how to configure the maximum index size and set limits for disk usage. You learn how to configure peer nodes in an offline state, enable maintenance mode, perform a rolling restart, remove bucket copies, and remove peer nodes. You learn what to do if the manager site fails. You go through REST API endpoint capabilities in Splunk and look at the Splunk SDK.

This chapter covers the following topics:

- Managing indexes

- Managing index storage

- Managing index clusters

- Managing a multisite index cluster

- Splunk REST API endpoints

- Splunk SDK

Managing Indexes

As you saw in earlier chapters, an index in Splunk is a logical collection of data stored on disk as a set of directories and files. These index directories are called buckets, and they're grouped by age—for example, hot, warm,

© Carlos Moreno Buitrago, Deep Mehta 2026

C. M. Buitrago and D. Mehta, *The Splunk Core User Study Companion*, Certification Study Companion Series, https://doi.org/10.1007/979-8-8688-2501-9_16

cold, and frozen (with thawed used when you restore archived data). By default, bucket directories live under `$SPLUNK_HOME/var/lib/splunk`. Splunk Enterprise provides support for two types of indexes:

- **Event indexes** are for general-purpose, optimized for event data. They can technically store almost any text-based data.

- **Metrics indexes** are used in a highly structured format to handle the higher volume and lower latency demands associated with metrics data.

Let's see how to configure event indexes and metrics indexes.

Configure Event Indexes

You can configure event indexes using Splunk Web, CLI, or a configuration file. If you want to configure your indexes **for the entire index cluster**, you need to deploy them using the **cluster manager**.

Configure Event Indexes Using a Splunk Configuration File

To add a new index for event indexes in the Splunk configuration file, go to `$SPLUNK_HOME/etc/system/local`, and edit indexes.conf or deploy a custom app. In the stanza name, provide the index name. You can refer to the following stanza block for more information:

```
[<index_name>]
homePath=<path_hot_warm>
coldPath=<path_cold>
thawedPath=<path_thawed>
```

- **homePath** defines <homePath> in configuring event indexes. Normally, data in warm and hot buckets resides in <homepath> in an index.

- **coldPath** defines <coldpath> in configuring event indexes. Normally, data in cold buckets resides in <coldPath> in an index.

- **thawedPath** defines <thawedpath> in configuring event indexes. Used for thawed buckets restored from archived frozen data.

Configure Event Indexes Using Splunk CLI

To configure event indexes using Splunk CLI, go to $SPLUNK_HOME/bin and refer to the following command to configure event indexes:

```
splunk add index <index_name> -homePath <homePath> -coldPath
<coldPath> -thawedPath <thawedPath>
```

Configure Metric Indexes Using a Splunk Configuration File

```
[<index_name>]
homePath=<path_hot_warm>
coldPath=<path_cold>
thawedPath=<path_thawed>
datatype=metric
```

- **datatype** defines the type of index, whether it is metric or event.

To configure metric indexes using Splunk CLI, go to $SPLUNK_HOME/bin and refer to the following command:

```
splunk add index <index_name> -datatype metric -homePath
<homePath> -coldPath <coldPath>  -thawedPath <thawedPath>
```

Remove Indexes and Index Data from Indexes

Removing indexes and index data includes removing data from the index, removing all the data from Splunk, removing the entire index, and even disabling it. This is possible using Splunk CLI commands.

- Refer to the following command to remove all buckets (data) from a specific index, but keep the index definition in indexes.conf:

  ```
  splunk clean eventdata -index <index_name>
  ```

- Refer to the following command to remove all data from all non-internal indexes (use only in labs or when rebuilding a system):

  ```
  splunk clean eventdata
  ```

- Refer to the following command to remove a particular index, including the index definition in indexes.conf:

  ```
  splunk remove index <index_name>
  ```

- Disabling an index is a better option than deleting an index. When you disable an index, you can re-enable it if you want:

  ```
  splunk disable index <index_name>
  ```

- To enable an index in Splunk, refer to the following command:

```
splunk enable index <index_name>
```

Configure Index Parallelization for Managing Indexes

Index parallelization is a feature that allows an indexer to maintain multiple pipeline sets. A *pipeline set* handles data processing from the ingestion of raw data, through event processing, and to writing the events to disk. An index runs a single pipeline set by default; however, if the instance is underutilized in terms of available CPU cores and I/O, you can configure the index to run multiple pipeline sets. By running multiple pipeline sets, you improve the index's indexing throughput capacity.

Configure Pipeline Sets for Index Parallelization

To configure pipeline sets for index parallelization, edit the server.conf if it exists, create a server.conf file at $SPLUNK_HOME/etc/system/local, or use a custom app with the following stanza (remember that for index clusters, the configuration should be deployed using the cluster manager):

```
[general]
parallelIngestionPipelines = 2
```

Configure the Index Allocation Method for Index Parallelization

As mentioned, when index parallelization is enabled, Splunk can run multiple ingestion pipeline sets in parallel. You can control how events are distributed across those pipeline sets by setting an allocation policy in the server.conf file located at $SPLUNK_HOME/etc/system/local or deploying a custom app.

There are two methods to configure allocation methods in Splunk: It can be either a round-robin selection method or a weighted-random selection method. Refer to the following stanza:

```
[general]
pipelineSetSelectionPolicy = round_robin | weighted_random
```

- **round_robin**: Splunk cycles through pipeline sets in order. This provides a simple, even distribution.

- **weighted_random**: Splunk selects pipeline sets using an internal weighting strategy, which can help balance work when some pipelines are busier than others.

In practice, round_robin is easier to reason about and is a safe default unless you have a specific reason to use weighted_random.

Manage Index Storage

Managing index storage is crucial in Splunk. You can configure index storage, move index, configure index size, set limits in index storage, and set retirement and archiving policies.

Define a New Index Database

You can change the default path of the index database in Splunk by changing the path definition of SPLUNK_DB. Keep in mind, it should be applied at the beginning of the indexer configuration. That change won't migrate automatically data from the old to the new path if the indexer has data already.

Follow these steps to use a new path:

1. Create the directory with write permissions.

2. Stop the indexer using Splunk CLI.

```
splunk stop
```

3. (Optional, to migrate the data to the new path) Copy
    the current index file system to the directory.

4. Unset the Splunk environment variable using
    system shell.

```
unset SPLUNK_DB
```

5. Change the SPLUNK_DB attribute in $SPLUNK_HOME/
    etc/splunk-launch.conf to the path you want.

```
SPLUNK_DB=<path>
```

6. Start the indexer using Splunk CLI.

```
splunk start
```

Configure Maximum Index Size for Indexer Storage

To configure the maximum index size for indexer storage in Splunk, you
can use the indexes.conf file. You learn how to set the maximum size for
each bucket and set the maximum index size for volume.

To configure the maximum size for an index in a non-cluster Splunk
indexer, modify indexes.conf in $SPLUNK_HOME/etc/system/local or
create a custom app, and add maxTotalDataSizeMB in the configuration
file. For an indexer cluster, just deploy a new custom app across the cluster.
Refer to the following code block for more help:

```
[<index_name>]
homePath.maxDataSizeMB = <value>
coldPath.maxDataSizeMB = <value>
```

- <index_name> is the name of the index.

- homePath.maxDataSizeMB is the maximum value of the bucket homepath, and the value is in megabytes (MB).

- coldPath.maxDataSizeMB stands for the maximum value of the bucket coldpath, and the value is in megabytes (MB).

To control how much data Splunk can store on a given volume, you define the volume in indexes.conf and set a size limit:

```
[volume:<volume_name>]
path = < filesystem_path>
maxVolumeDataSizeMB = <value>
```

- <volume_name> is the logical name of the volume you're defining.

- path is the file system path where the data for that volume will be stored.

- maxVolumeDataSizeMB is the maximum amount of data (in MB) that Splunk is allowed to store on this volume (across all indexes that use it).

To configure a volume for each index in Splunk, refer to the following stanza:

```
[<index_name>]
homePath = volume:<hotwarm_volume_name>/<index_name>
coldPath = volume:<cold_volume_name>/<index_name>
```

- In homePath and coldPath, volume:<hotwarm_volume_name>/volume:<cold_volume_name>, refer to the volume definitions you created in the [volume:<name>] stanzas.

- <index_name> is the name of the index, and it becomes the directory name under that volume's path.

Set Limit for Disk Usage in Splunk

To limit disk usage in Splunk, you can set the minimum amount of free space for the disk where the indexed data is stored. If the limit is reached, the indexer stops indexing data.

Configure Splunk to Set a Limit for Disk Usage Using Splunk CLI

The limit for disk usage is used mainly for disk space issues, so that you don't run out of disk space. To configure Splunk to limit disk usage in a non-cluster indexer, use Splunk CLI. Go to $SPLUNK_HOME/bin and refer to the following command:

```
splunk set minfreemb <size>
```

To enable configuration, you need to restart the Splunk instance.

Configure Splunk to Set a Limit for Disk Usage Using a Splunk Configuration File

To configure Splunk to set a limit for disk usage using a Splunk configuration file, edit the server.conf one. Refer to the following code block:

```
[diskUsage]
minFreeSpace = <MB>
```

minFreeSpace sets the limit for disk usage in Splunk. It is followed by a value that indicates how much free space the Splunk index should have.

Now let's move to the next section to learn best practices for index clusters.

Managing Index Cluster

Managing index clustering in Splunk comprises tasks like taking a peer node offline, using maintenance mode, a rolling restart of the index cluster, and removing excess buckets from the index cluster. Let's look at each topic, starting with configuring a peer node to be offline.

Configuring a Peer Node to Offline

To minimize search disruption in Splunk due to a maintenance window, you can take a peer down temporarily or permanently from your existing Splunk environment.

Configure Splunk to Offline Mode Using Splunk CLI

The Splunk offline command is used to gracefully remove an indexer peer from an indexer cluster. Without any flags, Splunk offline takes the peer offline temporarily for maintenance: the Cluster Manager redistributes primary buckets and triggers bucket fix-up to meet the replication factor (RF) and search factor (SF), while the peer waits for both primary reallocation and any active searches to complete.

These wait periods are controlled by timeout settings such as decommission_node_force_timeout and decommission_search_jobs_wait_secs on the peer. For long or slow-running clusters, you can override the decommission timeout directly (e.g., splunk offline --decommission_node_force_timeout <seconds>) and adjust the cluster-wide restart timeout with splunk edit cluster-config -restart_timeout <seconds>.

To remove a peer permanently from the cluster, use splunk offline --enforce-counts, which ensures that the cluster enforces its replication and search factors without that node before the shutdown completes:

- **splunk offline** takes a peer offline for maintenance purposes.

- **splunk offline --enforce-counts** permanently takes down a peer from the cluster.

To change the cluster restart timeout (how long the CM will wait for peers during its own restart), use the following command:

```
splunk edit cluster-config -restart_timeout <seconds>
```

Configure Splunk to Maintenance Mode Using Splunk CLI

Maintenance mode temporarily pauses the bucket fix-up activities in an indexer cluster and prevents hot buckets from rolling. This is useful when you perform maintenance on a peer (e.g., storage work or OS patching) and want to avoid extra cluster churn while the node is under stress. You should only enable maintenance mode when truly necessary and remember to turn it off when you're done.

To enable maintenance mode using Splunk CLI, refer to the following stanza:

```
splunk enable maintenance-mode
```

To return the standard bucket-rolling behavior to normal mode, refer to the following stanza:

```
splunk disable maintenance-mode
```

Rolling Restart in Splunk

A rolling restart performs a staged, coordinated restart of all cluster peers instead of stopping them all at once. The CM brings down a limited number of peers at a time (based on its configured restart slots), waits for them to come back and for RF/SF to be satisfied, and then moves on to the next set. This approach keeps searches running and ensures that load-balanced forwarders always have healthy indexers available to receive data.

Restarting all peers simultaneously is technically possible but not recommended, because it can interrupt ingestion and search. Instead, the CM controls how many peers can restart in parallel via its restart-slot settings, effectively limiting the percentage of the cluster that is allowed to be down at any given time.

Specify the Percentage of Peer to Restart at a Time Using Splunk CLI

The restart percentage can be configured via Splunk Web, a configuration file, or the CLI. To specify the percentage of peers to restart using Splunk CLI, refer to the command provided below:

```
splunk edit cluster-config -percent_peers_to_restart
<percentage>
```

Searchable Rolling Restart Using Splunk CLI

Splunk Enterprise 7.1.0 and later provides a searchable option for rolling restarts. The searchable option lets you perform a rolling restart of peer nodes with minimal interruption of ongoing searches. You can use a searchable rolling restart to minimize search disruption when a rolling restart is required due to regular maintenance or a configuration bundle push. To perform a searchable rolling restart, refer to the following command on the manager node:

```
splunk rolling-restart cluster-peers -searchable true
```

If you want to proceed with the searchable rolling restart despite the health check failure, use the force option on the manager node. It is not advisable because it may lead to clearing the queue data that is not indexed by the indexer. Refer to the following command:

```
splunk rolling-restart cluster-peers -searchable true
-force true \
-restart_inactivity_timeout <secs> \
-decommission_force_timeout <secs>
```

Remove Excess Bucket Copies from the Indexer Cluster

Excess bucket copies are bucket replicas that go beyond what the cluster's RF/SF requires. For example, if RF is 2 and a bucket ends up with three copies, that third copy is "excess." These extra copies often appear after peer outages, maintenance, or data rebalancing: while a peer is down, the CM creates additional copies to keep RF/SF met, and when the peer comes back, the temporary copies can remain.

To list extra bucket copies from a Splunk index cluster, refer to the following command:

```
splunk list excess-buckets [index_name]
```

To remove extra bucket copies from a Splunk index cluster, refer to the following command:

```
splunk remove excess-buckets [index_name]
```

Remove a Peer from Manager's List

After a peer goes down permanently, it remains on the manager node lists. For example, a peer goes down permanently, but it continues to appear on the manager dashboard, although its status changed to Down or Graceful Shutdown, depending on how it went down. In these cases, you need to remove the peer permanently from the manager node.

To remove a peer from the manager list using Splunk CLI, refer to the following stanza:

```
splunk remove cluster-peers -peers <guid>,<guid>,<guid>,...
```

Managing a Multisite Index Cluster

Managing multisite index clustering in Splunk consists of various tasks, like handling multisite failure, restarting indexing in a multisite cluster after a manager restart or site failure, converting a multisite index cluster to a single site, and moving a peer to a new site.

Manager Site in Multisite Index Cluster Fails

If the CM fails in a multisite indexer cluster, the cluster does not immediately stop working. The existing peer nodes continue to receive data from forwarders and serve searches based on the last known cluster state. However, while the CM is down, the cluster cannot enforce RF/SF, reassign primaries, or coordinate bucket fix-up and rolling maintenance. If additional peers fail during this period, the cluster cannot rebalance or protect data as designed.

For this reason, production multisite deployments should have a standby CM and a documented failover process. A standby CM lets you restore the control plane quickly so the cluster can resume enforcing RF/SF and managing replication across sites.

Configure Standby Server

To configure a standby server, you need to take care of the following
two things:

- Back up the files that the replacement manager needs.

- Ensure that the peer and search head nodes can find
  the new manager.

Back Up the Files That the Replacement Manager Needs

There are two static configurations that you must back up to copy to the
replacement manager:

- The manager's server.conf file, where the manager's
  cluster settings are stored.

- The manager's $SPLUNK_HOME/etc/manager-apps
  directory must be common where all peer node
  configuration is stored.

Ensure That the Peer and Search Head Nodes Can Find the New Manager

To ensure that the peer and search head nodes can find a new manager,
you need to follow any one rule from the given rules:

- The replacement must use the same IP address or DNS
  record and management port as the primary manager.

- If the replacement does not use the same IP address
  and management port as the primary manager, then
  configure the peer node and add a new manager_uri
  address.

Restart Indexing in the Multisite Cluster After a Manager Restart or Site Failure

When the CM restarts in a multisite cluster, it evaluates whether enough peers are available to satisfy the configured RF. If that condition is not met, the cluster marks itself as not indexing ready and blocks new indexing until additional peers come online. In special cases, such as a prolonged site outage, you can override this protection and forcibly resume indexing by running:

```
splunk set indexing-ready -auth admin:<password>
```

Move a Peer to a New Site

If you want to move a peer node from one site to another in a multisite environment, you can do it using the following instructions:

1. Take the peer offline with the offline command. The manager reassigns the bucket copies handled by this peer to other peers on the same site.

2. Ship the peer's server to the new site.

3. Delete the entire Splunk Enterprise installation from the server, including its index database and all bucket copies.

4. Reinstall Splunk Enterprise on the server, re-enable clustering, and set the peer's site value to the new site location.

Let's now discuss how to use REST API endpoints in Splunk.

REST API Endpoints

Splunk REST API endpoints can do almost all operations in Splunk, from authentication to searching to configuration management. The API is divided into endpoints (URIs) served by the splunkd process (i.e., management port 8089). REST API endpoints can be used in Splunk by a user for remote querying, searching remotely, and using a third-party to integrate their apps with Splunk. Splunk provides a Software Development Kit (SDK) for developers to integrate their app with Splunk. The SDK is like a wrapper that calls the REST API and helps abstract the details by providing easy-to-use objects that can interact with Splunk.

In REST API endpoints, you use the open source command-line tool, CURL. There are other command-line tools available, such as wget. CURL is available on Mac and Linux by default. It can also be downloaded for Windows; go to `http://curl.haxx.se/` for more information.

There are three main methods in Splunk REST API endpoints:

- The GET method gets data that is associated with a resource; for example, accessing search for a result.

- The POST method creates or updates an existing resource.

- The DELETE method deletes a resource.

REST API endpoints have three main functions:

- Running searches

- Managing knowledge objects and configuration

- Updating Splunk Enterprise configuration

Running Searches Using REST API

When you want to run a search in Splunk, you are really doing three things:

- Create a search job.

- Poll its status.

- Fetch the results.

Create a Search Job

When you create the job, you can pass several useful parameters:

- **max_count:** Set this parameter if the search result is greater than 10,000 events.

- **status_buckets**: To access a summary and timeline information from a search job, specify a value for this parameter.

- **rf** (required fields): Use this parameter to **force specific fields to be included** in the results/summary for the job, even if they're not part of the default field set (you can specify it multiple times).

To create a search job in Splunk, follow these steps:

1. Open the terminal and go to $SPLUNK_HOME/bin.

2. Execute the curl command. You can use the following example:

```
curl -u admin:<password> -k https://localhost:8089/
services/search/jobs    \
-d search="search index=_internal | head 10" \
-d max_count=50000 \
-d status_buckets=300
```

3. Splunk returns XML with a search ID (SID). By
    default, SID is valid for ten minutes. The following
    code block is the reply from Splunk:

```
<?xml version='1.0' encoding='UTF-8'?>
<response>
   <sid>1768421821.56</sid>
</response>
```

4. To poll the job to see if it's done (status), type the
    following command:

```
curl -u admin:<password> -k https://localhost:8089/
services/search/jobs/1768421821.56
```

    If you want to know whether your search was
    successful or not, in reply, you get a message
    Job Done.

5. To get the results of your search operation, execute
    the following command:

```
curl -u admin:<password> -k https://localhost:8089/
services/search/jobs/1768421821.56/results
```

6. To get the result in CSV or JSON format, refer to the
    following command:

```
curl -u admin:<password> -k https://localhost:8089/
services/search/jobs/1768421821.56/results --get -d
output_mode=<csv/json>
```

Manage Configurations File in Splunk

Two sets of endpoints provide access to the configuration files:

- **properties/** endpoint is convenient for **browsing and editing key/value pairs** in configuration files.

- **configs/conf-<file>/** endpoint used to **manage stanzas (entries) and ACL/permissions** for a given configuration file.

Together, they let you inspect and update *.conf files directly from scripts or automation, without manually editing files on disk.

Example: list stanzas and settings in props.conf in the search app context:

```
curl -k -u admin:<password> https://localhost:8089/servicesNS/
nobody/search/properties/props
```

Example: add a new stanza to props.conf in the search app:

```
curl -k -u admin:<password> https://localhost:8089/servicesNS/
nobody/search/configs/conf-props \
  -d name=test89 \
-d SHOULD_LINEMERGE=false
```

This brings us to the end of Splunk REST API endpoints. Now let's move to the Splunk SDK.

Splunk SDK

In this section, you learn about using Splunk's Software Development Kit (SDK), which is available in Python, Java, JavaScript, and C#. Generally, since the Python SDK is used most often, it is covered in this book. At a high level, the SDKs are thin wrappers around the REST API and let you

- Run searches and saved searches from external programs.

- Manage Splunk objects (indexes, inputs, knowledge objects, apps, etc.).

- Push data directly into Splunk (e.g., from a script or service).

- Export data from Splunk into other systems for long-term storage or further processing.

- Build custom tools, dashboards, and integrations without manually crafting HTTP requests.

Python Software Development Kit for Splunk

The Splunk SDK for Python helps developers interact with Splunk for various operations, including searching, saved searches, data input, REST API endpoints, building applications, and so forth. The SDK supports modern Python 3.x and is distributed as a standard Python package. You can either

- Install it from its GitHub repository.

- Install it as a library in your Python environment and then import it in your scripts.

The official source and documentation are available at `https://github.com/splunk/splunk-sdk-python`.

If you're working from a cloned copy of the SDK rather than an installed package, you may need to adjust your PYTHONPATH so that Python can find the Splunk SDK modules.

Program for Data Input in Splunk Using splunklib.client

To work with an SDK using splunklib.client, you need not configure the
.splunkrc file; have the splunklib.client file in your Python environment.
The following is a simple program showing how to push data from a
particular file to the main index in your Splunk environment:

```python
import splunklib.client as client
HOST = "localhost"
PORT = 8089
USERNAME = "admin"
PASSWORD = "Deep@1234"
service = client.connect(
    host=HOST,
    port=PORT,
    username=USERNAME,
    password=PASSWORD)
myindex = service.indexes["main"]
file=open("/Applications/Splunk/copy.txt", "w")
file.write("hi")
file.close()
uploadme = "/Applications/Splunk/copy.txt"
myindex.upload(uploadme);
print(myindex.upload(uploadme))
print("successful")
```

After executing the program, go to your Splunk Web environment and
check out the output. It is similar to Figure 16-1.

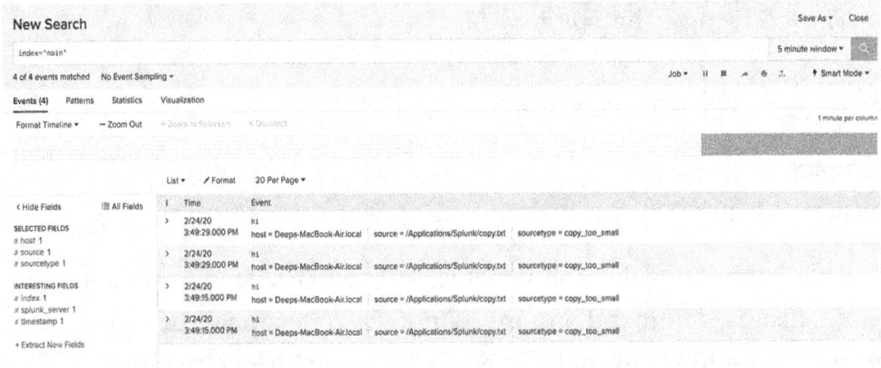

Figure 16-1. *Output "Search.py" events using Splunk Web*

Program for Search in Splunk Using a Command Line

To work with the examples provided within the Splunk SDK, you need to configure the .splunkrc file in your respective environment at `http://dev.splunk.com/view/python-sdk/SP-CAAAEFC`.

The following is the .splunkrc file configuration syntax:

```
host=<ip>,<hostname>,<localhost>
port=<mgmt port>
username=<username>
password=<password>
```

- **host** is the address on which you have access to the Splunk instance.

- **port** refers to your mgmt_port (8089 by default).

- **username** is the username of Splunk instance (admin by default).

- **password** secures your account.

The following is the .splunkrc file:

```
host=localhost
port=8089
username=admin
password=Deep@1234
```

After configuring the .splunkrc file, try to run a search.py program in Python using a command prompt (Windows) or terminal (Linux). Go to the example splunk-sdk folder and run the following command:

```
python search.py " search index=main|head 1"
```

The output of the search.py program is shown in Figure 16-2.

```
(base) bash-3.2$ python search.py " search index=main|head 1"
<?xml version='1.0' encoding='UTF-8'?>
<results preview='0'>
<meta>
<fieldOrder>
<field>_bkt</field>
<field>_cd</field>
<field>_indextime</field>
<field>_raw</field>
<field>_serial</field>
<field>_si</field>
<field>_sourcetype</field>
<field>_time</field>
<field>host</field>
<field>index</field>
<field>linecount</field>
<field>source</field>
<field>sourcetype</field>
<field>splunk_server</field>
</fieldOrder>
</meta>
        <result offset='0'>
                <field k='_bkt'>
                        <value><text>main~0~EB365685-8BB0-4299-BE30-24A1982CB10C</text></value>
                </field>
                <field k='_cd'>
                        <value><text>0:12</text></value>
                </field>
                <field k='_indextime'>
                        <value><text>1582541220</text></value>
                </field>
                <field k='_raw'><v xml:space='preserve' trunc='0'>hi</v></field>
                <field k='_serial'>
                        <value><text>0</text></value>
                </field>
                <field k='_si'>
                        <value><text>Deeps-MacBook-Air.local</text></value>
                        <value><text>main</text></value>
                </field>
                <field k='_sourcetype'>
                        <value><text>copy-too_small</text></value>
                </field>
                <field k='_time'>
                        <value><text>2020-02-24T16:17:00.000+05:30</text></value>
                </field>
                <field k='host'>
                        <value><text>Deeps-MacBook-Air.local</text></value>
                </field>
                <field k='index'>
                        <value><text>main</text></value>
                </field>
                <field k='linecount'>
                        <value><text>1</text></value>
                </field>
                <field k='source'>
                        <value><text>/Applications/Splunk/copy.txt</text></value>
                </field>
                <field k='sourcetype'>
                        <value><text>copy-too_small</text></value>
                </field>
                <field k='splunk_server'>
                        <value><text>Deeps-MacBook-Air.local</text></value>
```

Figure 16-2. *Output "search.py" events using terminal*

You can run other example .py files and go through the reference material to become more familiar with the Splunk SDK and with the Splunk environment.

Summary

In this journey of indexes, you learned how to configure custom indexes, remove indexes, index data, and configure index parallelization. You saw how to configure the maximum index size and set limits for disk usage. You learned how to configure a node to be in an offline state, enable maintenance mode, rolling restart, remove bucket copies, and remove peer nodes. You learned about Splunk REST API endpoints, the Splunk SDK, and the Python SDK and executed a test program.

This chapter covered the following topics of the Splunk Enterprise Certified Architect exam blueprint:

- **Module 7**: Performance Monitoring and Tuning—5%

- **Module 15**: Single-Site Index Cluster—5%

- **Module 16**: Multisite Index Cluster—5%

- **Module 17**: Indexer Cluster Management and Administration—7%

Multiple-Choice Questions

A. A Splunk admin wants to configure pipeline sets for index parallelization. Select the correct option from the following.

1. [setup]

   parallelIngestionPipelines = 2

2. [setup]

   parallelIngestionPipellines = 2

3. [general]

   parallelIngestionPipelines = 2

4. [general]
    doubleIngestion = true

B. The Splunk Software Development Kit is available in
    which languages? (Select all options that apply.)

1. Python

2. Java

3. JavaScript

4. Impala

5. Scala

6. None of the above

C. Which methods are for REST API endpoints
    in Splunk?

1. GET

2. POST

3. DELETE

4. INSERT

5. None of the above

D. Which is the default management port in Splunk?

1. 8000

2. 8089

3. 8001

4. 8098

    E. To configure event indexes, which file do you need to edit?

       1. input.conf

       2. deployment.conf

       3. output.conf

       4. indexes.conf

Answers

    A. 3

    B. 1, 2, 3

    C. 1, 2, 3

    D. 2

    E. 4

References

- https://help.splunk.com/en/splunk-enterprise/get-started/deployment-capacity-manual/10.0/performance-reference/parallelization-settings
- https://help.splunk.com/en/splunk-enterprise/administer/manage-indexers-and-indexer-clusters/10.0/manage-the-indexer-cluster/perform-a-rolling-restart-of-an-indexer-cluster

- https://help.splunk.com/en/data-management/
  manage-splunk-enterprise-indexers/10.0/manage-
  the-indexer-cluster/remove-excess-bucket-
  copies-from-the-indexer-cluster

- https://help.splunk.com/en/data-management/
  manage-splunk-enterprise-indexers/10.0/manage-
  a-multisite-indexer-cluster/restart-indexing-
  in-multisite-cluster-after-manager-restart-or-
  site-failure

- https://help.splunk.com/en/data-management/
  manage-splunk-enterprise-indexers/10.0/manage-
  a-multisite-indexer-cluster/move-a-peer-to-
  a-new-site

- https://dev.splunk.com/enterprise/docs/python/
  sdk-python/examplespython/commandline

- https://docs.splunk.com/DocumentationStatic/
  PythonSDK/1.6.5/client.html

- https://help.splunk.com/en/splunk-enterprise/
  leverage-rest-apis/rest-api-reference/10.0/
  introduction/endpoints-reference-list

- https://help.splunk.com/en/splunk-enterprise/
  administer/manage-indexers-and-indexer-
  clusters/10.0/manage-indexes/create-
  custom-indexes

- https://help.splunk.com/en/splunk-enterprise/
  administer/manage-indexers-and-indexer-
  clusters/10.0/manage-indexes/create-
  custom-indexes

CHAPTER 17

Final Practice Set

This chapter features multiple-choice questions that are useful for Splunk Enterprise Certified Admin and Architect certification exams. You get a better idea of the types of questions that appear on these exams. Keep in mind that the Splunk Enterprise Certified Architect exam requires you to complete four courses before taking the exam:

- Architecting Splunk Enterprise Deployments

- Troubleshooting Splunk Enterprise

- Splunk Cluster Administration

- Splunk Enterprise Deployment Practical Lab

Questions

A. **In a multisite index cluster, which configuration stores two searchable copies at the origin site, one searchable copy at site2, and a total of four searchable copies?**

1. site_search_factor = origin:2, site1:2, total:4

2. site_search_factor = origin:2, site2:1, total:4

3. site_replication_factor = origin:2, site1:2, total:4

4. site_replication_factor = origin:2, site2:1, total:4

© Carlos Moreno Buitrago, Deep Mehta 2026

C. M. Buitrago and D. Mehta, *The Splunk Core User Study Companion*, Certification Study Companion Series, https://doi.org/10.1007/979-8-8688-2501-9_17

B. **If you suspect that there is a problem interpreting a regular expression in a monitor stanza, which log file would you search to verify?**

1. btool.log

2. metrics.log

3. splunkd.log

4. tailing_processor.log

C. **When should multiple pipelines be enabled?**

1. Only if the disk IOPS is at 800 or better

2. Only if there are fewer than 12 concurrent users

3. Only if running Splunk Enterprise version 6.6 or later

4. Only if CPU and memory resources are significantly under-utilized

D. **Which tool(s) can be leveraged to diagnose connection problems between an indexer and forwarder? (Select all that apply.)**

1. Telnet

2. tcpdump

3. Splunk btool

4. Splunk btprobe

E. **Which CLI command converts a Splunk instance to a license peer?**

1. splunk add licenses

2. splunk list licenser-slaves

3. splunk edit licenser-localpeer -manager_uri <manager:port>

4. splunk list licenser-localslave

F. **Which Splunk server role regulates the functioning of an indexer cluster?**

1. Indexer

2. Deployer

3. Manager node

4. Monitoring Console

G. **To improve Splunk performance, the parallelIngestionPipelines setting can be adjusted on which of the following components in the Splunk architecture? (Select all that apply.)**

1. Indexers

2. Forwarders

3. Search head

4. Cluster manager

H. **Of the following types of files within an index bucket, which file type may consume the most disk?**

1. Raw data

2. Bloom filter

3. Metadata (.data)

4. Inverted index (.tsidx)

I. **Which server.conf attribute should be added to the manager node's server.conf file when decommissioning a site in an indexer cluster?**

1. site_mappings

2. available_sites

3. site_search_factor

4. site_replication_factor

J. **Which two sections can be expanded using the search job inspector?**

1. Execution costs

2. Saved search history

3. Search job properties

4. Optimization suggestions

K. **What does the deployer do in a search head cluster (SHC)? (Select all that apply.)**

1. Distributes apps to SHC members

2. Bootstraps a clean Splunk install for an SHC

3. Distributes non-search-related and manual configuration file changes

4. Distributes runtime knowledge object changes made by users across the SHC

L. **When Splunk indexes data, what kind of files does it create by default? (Select all that apply.)**

1. Compress data

2. .tsidx files

3. props.conf and transforms.conf

4. Key/value database

M. **To activate replication for an index in an indexer cluster, what attribute must be configured in indexes.conf on all peer nodes?**

1. repFactor = 0

2. replicate = 0

3. repFactor = auto

4. replicate = auto

N. **Which search head cluster component is responsible for pushing knowledge bundles to search peers, replicating configuration changes to search head cluster members, and scheduling jobs across the search head cluster?**

1. Manager

2. Captain

3. Deployer

4. Deployment server

O. **Which Splunk internal index contains license-related events?**

1. _audit

2. _license

3. _internal

4. _introspection

P. **What is the default log size for Splunk internal logs?**

   1. 10 MB

   2. 20 MB

   3. 25 MB

   4. 30 MB

Q. **When Splunk is installed, where are the internal indexes stored by default?**

   1. $SPLUNK_HOME/bin

   2. $SPLUNK_HOME/var/run/splunk

   3. $SPLUNK_HOME/var/lib/splunk

   4. $SPLUNK_HOME/etc/system/default

R. **In which phase of the Splunk Enterprise data pipeline are indexed extraction configurations processed?**

   1. Input

   2. Search

   3. Parsing

   4. Indexing

S. **When planning a search head cluster, which of the following is true?**

   1. All search heads must use the same operating system.

   2. All search heads must be members of the cluster (no standalone search heads).

   3. The search head captain must be assigned to the largest search head in the cluster.

4. All indexers must belong to the underlying indexer cluster (no standalone indexers).

Answers

A. 2

B. 3

C. 4

D. 1, 2, 3

E. 3

F. 3

G. 1, 2

H. 4

I. 1

J. 1, 3

K. 1, 3

L. 1, 2

M. 3

N. 2

O. 3

P. 3

Q. 3

R. 3

S. 1

CHAPTER 18

Setting Up a Splunk Environment with AWS

In this chapter, you will put Splunk into a real cloud context by deploying it on **Amazon Web Services (AWS)**. You will build and configure Splunk on EC2 and then expand into a full multisite, clustered architecture. By the end, you'll be able to deploy and manage apps through the cluster manager and monitor the whole environment as a single, distributed system.

This chapter covers the following topics:

- A quick introduction to Amazon Web Services (AWS)

- Installing and configuring Splunk on Amazon EC2

- Deploying a multisite indexer cluster

- Deploying a search head cluster (SHC)

- Distributing configuration bundles through the cluster manager

- Monitoring and troubleshooting a distributed Splunk environment

© Carlos Moreno Buitrago, Deep Mehta 2026
C. M. Buitrago and D. Mehta, *The Splunk Core User Study Companion*, Certification Study
Companion Series, https://doi.org/10.1007/979-8-8688-2501-9_18

Amazon Web Services

Amazon Web Services (AWS) began offering public cloud services in 2006, and cloud offerings are commonly described using three models: Infrastructure as a Service (IaaS), Platform as a Service (PaaS), and Software as a Service (SaaS).

IaaS provides the foundational building blocks, such as virtual compute, storage, and networking, that customers assemble and manage themselves. On AWS, typical IaaS examples include EC2, EBS, EFS, S3/Glacier, VPC, Elastic Load Balancing, Route 53, CloudFront, and Direct Connect.

PaaS sits one level higher by offering managed platforms and runtimes so customers can deploy applications without managing underlying servers; AWS examples include RDS, DynamoDB, Elastic Beanstalk, ECS/Fargate, EMR/Glue, and Lambda (serverless/FaaS).

SaaS delivers complete applications directly to end users, where AWS handles the full stack; AWS examples are WorkMail, WorkDocs, and WorkSpaces.

In this chapter, we use Amazon EC2 instances to deploy Splunk in AWS.

Configuring an EC2 Instance Using the AWS Management Console

Amazon EC2 instances can be configured in many ways, but the easiest and best way is to go to `https://aws.amazon.com/` and sign in to the console. Once you have signed in to your AWS account, you gain access to the Amazon Management Console page. You can manage your entire account from this page. Let's have a look at it (see Figure 18-1).

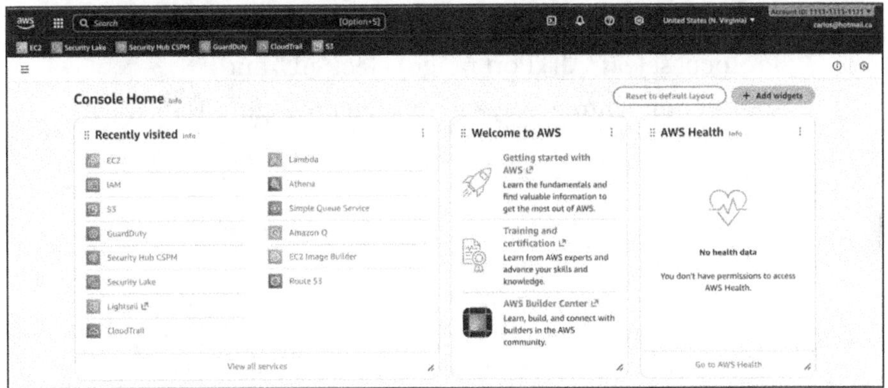

Figure 18-1. *AWS management console*

In this section, you use only EC2 instances to configure Splunk instances. In AWS, search for EC2 instances and then go to the EC2 page.

1. On the left side of the EC2 instance, there is the Network & Security menu bar.

2. Go to Key pairs.

3. Add a key (I created a key named *test*). Refer to Figure 18-2.

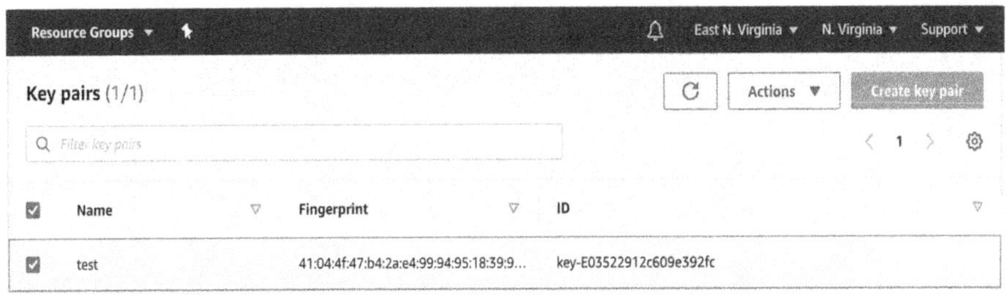

Figure 18-2. *Test key*

4. In Network & Security, go to Security Groups.

5. Create a new security group (I created a security group named Main).

6. Edit Inbound rules where in traffic select "All traffic",
   Protocol select "All", Port range select "All", and
   destination Source "Anywhere." Refer to Figure 18-3.

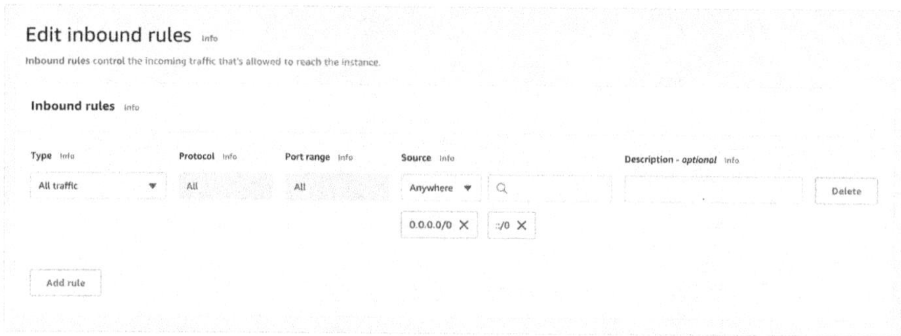

Figure 18-3. *Main security group*

7. Click the Launch instance in EC2 dashboards.

8. Select an image (whichever you want). I selected
   Amazon Linux 2023 AMI for this exercise. Click
   Next. Instance type provides a wide selection of
   instance types optimized for use cases (I selected
   t2.micro for this exercise).

9. Click Review and Launch.

10. Click Edit Security Groups ➤ Existing Security
    Groups, and then select the group you have created
    (Main for me).

11. Click Launch and select the key pair that you
    created. (I created a key named Test.)

In this chapter, you deploy Splunk multisite clustering for site A and
site B. Each site has two indexers, search head clustering, and a license
manager, a cluster manager, and a deployment server/Monitoring Console

on the same instance. Additionally, there will be two universal forwarders. The private IP of the instance is also provided. Refer to the diagram shown in Figure 18-4.

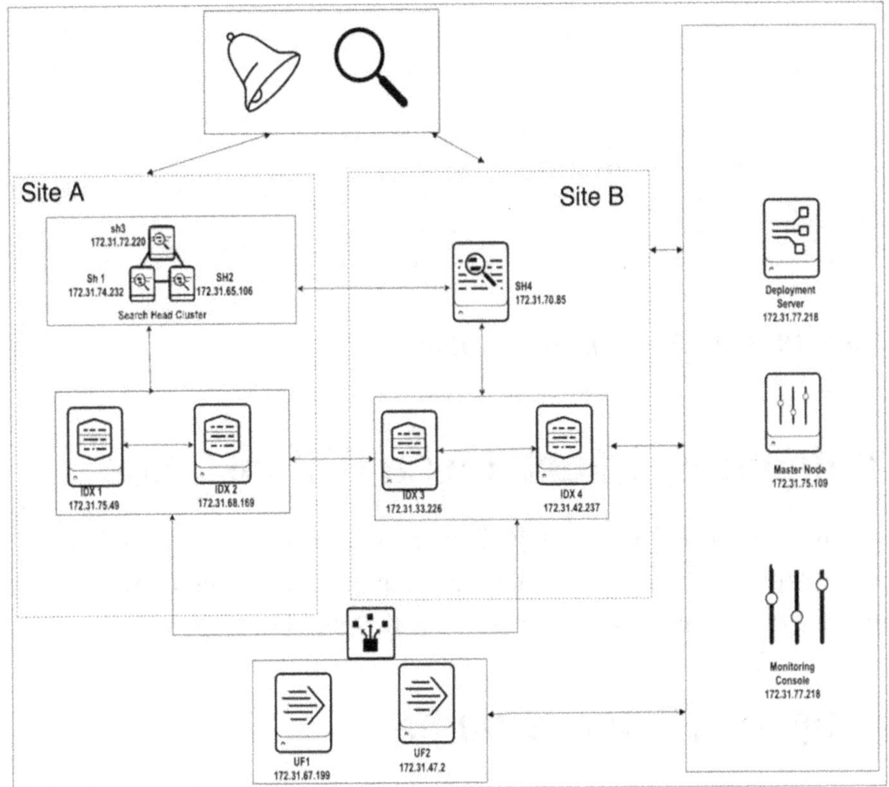

Figure 18-4. *Splunk deployment constituent architecture*

To deploy this multisite clustering example, you will need to have 12 different Splunk instances. Create EC2 instances as warranted, and keep in mind this is just a test exercise to implement multisite index clustering. For production deployment, it is not recommended at all due to the size of the instances and the use case we want to achieve. You can refer to Figure 18-5 for it.

Name	Instance ID	Instance Type	Availability Zone	Instance State	Status Checks	Alarm Status	Public IPv4 DNS	IPv4 Public IP	IPv6 IPs	Key Name	Monitoring
idx4	i-0df571c2d46d0a1d7	t3.micro	us-east-1a	running	2/2 checks …	None	ec2-3-87-25-60.comp…	3.87.25.60	-	test	disabled
sh2	i-0b277f1b2e3a4b5c6	t3.micro	us-east-1a	running	2/2 checks …	None	ec2-54-87-21-72.co…	54.87.21.72	-	test	disabled
ds/mc	i-04a97e7f8c8b3d2a1	t3.micro	us-east-1a	running	2/2 checks …	None	ec2-44-193-67-171.c…	44.193.67.171	-	test	disabled
idx1	i-0a1b2c3d4e5f67890	t3.micro	us-east-1a	running	2/2 checks …	None	ec2-34-201-90-2.co…	34.201.90.2	-	test	disabled
sh1	i-0123456789abcdef0	t3.micro	us-east-1a	running	2/2 checks …	None	ec2-54-209-79-15.co…	54.209.79.15	-	test	disabled
search_captain	i-09f6e7d6c5b4a3d21	t3.micro	us-east-1a	running	2/2 checks …	None	ec2-23-21-28-136.co…	23.21.28.136	-	test	disabled
idx3	i-0c1d2e3f4a5b6c7d8	t3.micro	us-east-1a	running	2/2 checks …	None	ec2-52-21-20-78.co…	52.21.20.78	-	test	disabled
idx2	i-0d4c3b2a1f0e9d8c7	t3.micro	us-east-1a	running	2/2 checks …	None	ec2-34-207-33-152.c…	34.207.33.152	-	test	disabled
license_mas..	i-0e5f6a7b8c9d0e1f2	t3.micro	us-east-1a	running	2/2 checks …	None	ec2-3-91-217-175.co…	3.91.217.175	-	test	disabled
uf1	i-0f1e2d3c4b5a69788	t3.micro	us-east-1a	running	2/2 checks …	None	ec2-52-27-228-96.co…	52.27.228.96	-	test	disabled
uf2	i-01a2b3c4d5e6f7a89	t3.micro	us-east-1a	running	2/2 checks …	None	ec2-54-215-92-75.co…	54.215.92.75	-	test	disabled
cmaster	i-02b3c4d5e6f7a8b90	t3.micro	us-east-1a	running	2/2 checks …	None	ec2-34-207-243-175…	34.207.243.175	-	test	disabled

Figure 18-5. *EC2 instances for Splunk*

Configuring Splunk on an EC2 Instance

You have created 12 Amazon Web Services EC2 instances for a multisite
Splunk deployment. In this deployment, you need to configure two Splunk
universal forwarders and ten instances with Splunk Enterprise.

Configuring Splunk Enterprise

To configure a Splunk Enterprise instance on Amazon Web Services EC2,
log in to the EC2 instance using your keys and then refer to the following
instructions:

1. Create a directory in /opt named **splunk**. We will
    use sudo during the process to make it easier, but it
    is recommended to follow the security guidelines,
    running Splunk with a dedicated user and set up the
    required permissions.

    ```
    $ sudo mkdir /opt/splunk
    ```

2. Download Splunk Enterprise. You can use the .tgz, .deb, or .rpm packages. We will use the .tgz one.

3. Extract the Splunk image in /opt. The following is an example:

```
$ sudo  tar xzvf <splunk_package_name>.tgz -C /opt
```

4. Install Splunk Enterprise and accept the license.

```
$ sudo  /opt/splunk/bin/splunk start --accept-license
```

5. The installation process requires setting the username and password.

6. Set the Splunk server name and hostname based on the instance type; for example, if it's indexer 1, name it idx1.

```
$ sudo  /opt/splunk/bin/splunk set servername idx1
```

7. We have to make sure Splunk will start after a server reboot and/or a hard stop.

```
$ sudo /opt/splunk/bin/splunk  enable boot-start
```

8. For the indexers, enable the Splunk port of your choice. It can be deployed with the cluster manager as well.

```
$ sudo ./splunk  enable listen 9997
```

Configuring Splunk Forwarder

Let's configure a Splunk forwarder on Amazon Web Services EC2 instances. Log in to EC2 instances and refer to the following instructions:

1. Create a directory in /opt named splunkforwarder.

    ```
    $ sudo mkdir /opt/splunkforwarder
    ```

2. Download Splunk Universal Forwarder. You can
    use the .tgz, .deb, or .rpm packages. We will use the
    .tgz one.

3. Extract a Splunk package in /opt. The following is an
    example:

    ```
    $ sudo  tar xzvf <splunk_package_name>.tgz -C /opt
    ```

4. Install Splunk and accept the Splunk license.

    ```
    $ sudo  /opt/splunkforwarder/bin/splunk
    start --accept-license
    ```

5. Set the username and password of your choice.

6. Set the Splunk server name and Splunk hostname
    based on instance type; for example, if it's forwarder
    1, name it uf1.

    ```
    $ sudo /opt/splunkforwarder/splunk set servername uf1
    $ sudo /opt/splunkforwarder/splunk set default-
    hostname uf1
    ```

7. Enable Splunk boot start.

    ```
    $ sudo /opt/splunkforwarder/splunk  enable boot-start
    ```

Next, let's move to Splunk multisite index clustering to configure a
manager site and peer nodes.

Deploying Multisite Index Clustering

In this section, you will configure multisite index clustering in Splunk where you have two sites, Site 1 (A) and Site 2 (B) (refer to Figure 18-4). You will deploy multisite index clustering where replication_factor=1, searchable_copies=1, and total=2.

Configuring a Cluster Manager

To configure a cluster manager in Splunk, log in to an EC2 instance for a cluster manager, install Splunk Enterprise as shown in previous topics, and go to $SPLUNK_HOME/bin/.

To enable a cluster manager in Splunk, refer to the following command, which will have searchable copies of 1 and replication factor of 1 per site:

```
$ sudo ./splunk edit cluster-config \
-mode manager \
-multisite true \
-site site1 \
-available_sites site1,site2 \
-site_replication_factor origin:1,total:2 \
-site_search_factor origin:1,total:2 \
-replication_factor 2 \
-search_factor 2 \
-secret idxcluster
```

To monitor the cluster manager's status, navigate to Splunk Web's settings and go to Indexer Clustering, where you find a screen similar to Figure 18-6.

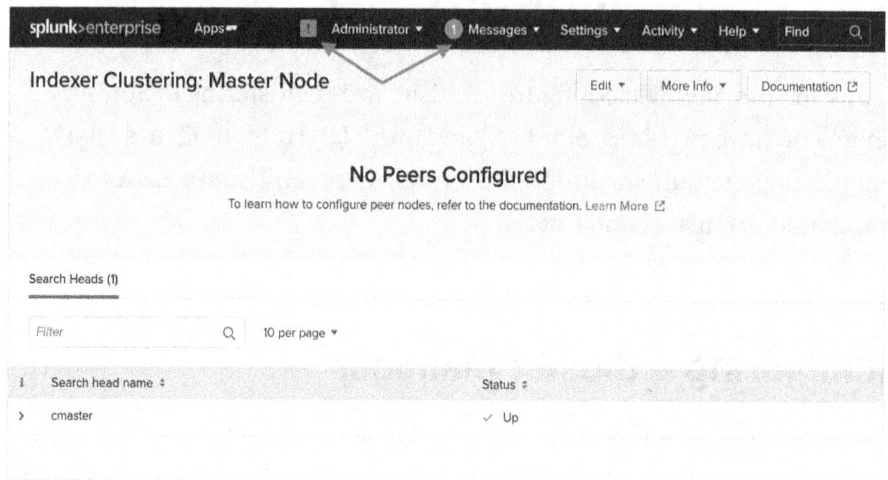

Figure 18-6. Monitoring Indexer Clustering manager node

Configuring a Peer Node

To configure a peer node for a cluster manager in Splunk, log in to the EC2 instance of an indexer and go to $SPLUNK_HOME/bin/.

To enable indexer 1 and indexer 2, report to the manager node with site 1. Refer to the following command:

```
$ sudo ./splunk edit cluster-config \
-mode peer \
-manager_uri https://172.31.75.109:8089 \
-replication_port 9887 \
-site site1 \
-secret idxcluster

 $ sudo ./splunk restart
```

To enable clustering in indexer 3 and indexer 4, run the following command:

```
$ sudo ./splunk edit cluster-config \
-mode peer \
-manager_uri https://172.31.75.109:8089 \
-replication_port 9887 \
-site site2 \
-secret idxcluster

$ sudo ./splunk restart
```

Configure the manager node and all indexers to their respective sites. To start monitoring the cluster, move to the cluster manager and navigate to Splunk Web's settings. Go to Indexer Clustering, where you find all the Splunk instances reported to the cluster manager. Figure 18-7 shows that idx1, idx2, idx3, and idx4 reported to the manager node.

Peer Name	Site	Fully Searchable	State	Buckets
idx3	site2	✓ Yes	Up	8
idx1	site2	✓ Yes	Up	8
idx2	site1	✓ Yes	Up	8
idx4	site1	✓ Yes	Up	8

Figure 18-7. *Monitoring Indexer Clustering peer nodes*

Next, let's move to deploying search head clustering.

Deploying a Search Head

In this section, you configure multisite search head clustering for site 1 and deploy a search head for site 2.

Configuring a Search Head

To configure a peer node for a search head in Splunk, log in to the EC2 instance of a search head and go to $SPLUNK_HOME/bin/.

To enable sh1, sh2, and sh3 and report to a manager node with site 1, refer to the following command:

```
$ sudo ./splunk edit cluster-config \
-mode searchhead \
-manager_uri https://172.31.75.109:8089 \
-site site1   \
-secret idxcluster

$ sudo ./splunk restart
```

To enable sh4 and report to a manager node with site 2, refer to the following command:

```
$ sudo ./splunk edit cluster-config \
-mode searchhead \
-manager_uri https://172.31.75.109:8089 \
-site site2   \
-secret idxcluster

$ sudo ./splunk restart
```

After assigning each search head to its proper site, you should confirm that the cluster manager (CM) has received their registrations. In Splunk Web, go to Settings ➤ Indexer Clustering, then open the Search Heads section. This view lists every search head that has successfully reported to the CM. As shown in Figure 18-8, the CM sees Deep-SHC (sh3) along with sh1, sh2, sh4, and cmaster as active reporting members.

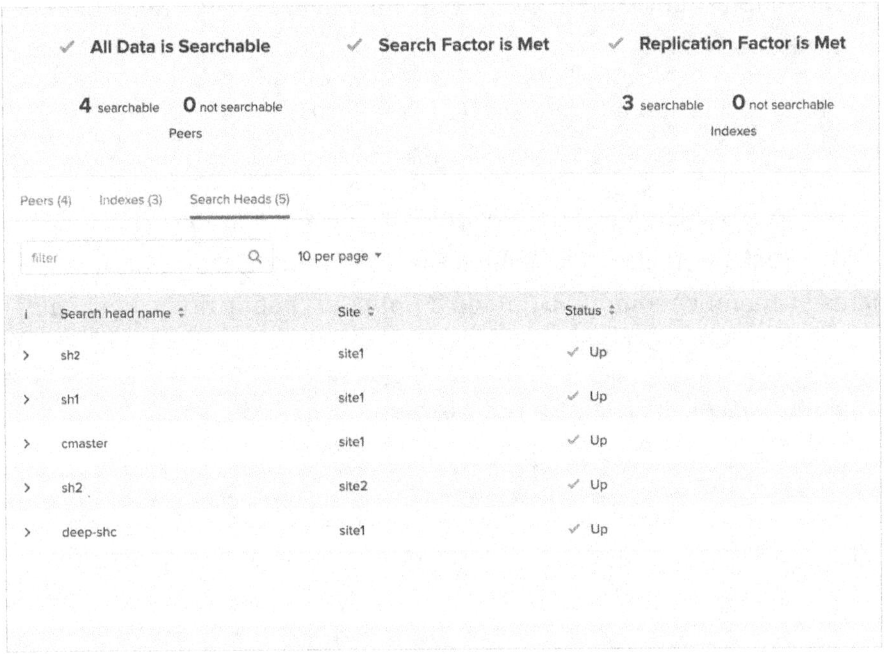

Figure 18-8. *Monitoring Indexer Clustering search heads*

Configuring Search Head Clustering

In this section, you would configure Splunk search head clustering for site 1 (i.e., sh1, sh2, and sh3 (Deep-shc)). To enable SHs for clustering with a secret key, use the init command on all the SHs. Refer to the following command:

```
$ sudo ./splunk init shcluster-config \
-mgmt_uri https://<sh_ip>:8089 \
-replication_port 9200 \
-secret shcluster

$ sudo ./splunk restart
```

Go to the instance where you want to make the node a search head captain; in my case, it's a Deep-shc node. Use the bootstrap command to elect a captain since there is only one node in the cluster. Refer to the following command.

```
$ sudo ./splunk bootstrap shcluster-captain \
-servers_list https://172.31.72.220:8089
```

This forces sh3 to become captain so the cluster can start forming.

To add sh1 to the existing Splunk search head cluster, use the add shcluster-member command, but do it from the captain node only. Refer to the following command:

```
$ sudo ./splunk add shcluster-member \
-new_member_uri https://172.31.74.232:8089
```

To add sh2 to the existing Splunk search head cluster, use the add shcluster-member command, but do it from the captain node only. Refer to the following command:

```
$ sudo ./splunk add shcluster-member \
-new_member_uri https://172.31.65.106:8089
```

```
Captain:
                          dynamic_captain : 1
                          elected_captain : Tue Jun 23 13:16:02 2020
                                       id : F5C922BF-95C7-4D41-BDB8-8DDFAC47ABD7
                         initialized_flag : 1
                                    label : Deep-SHC
                                 mgmt_uri : https://172.31.72.220:8089
                  min_peers_joined_flag : 1
                     rolling_restart_flag : 0
                      service_ready_flag : 1


Members:
    sh2
                                    label : sh2
                    last_conf_replication : Tue Jun 23 13:18:30 2020
                                 mgmt_uri : https://172.31.65.106:8089
                           mgmt_uri_alias : https://172.31.65.106:8089
                                   status : Up
    sh1
                                    label : sh1
                    last_conf_replication : Tue Jun 23 13:18:32 2020
                                 mgmt_uri : https://172.31.74.232:8089
                           mgmt_uri_alias : https://172.31.74.232:8089
                                   status : Up
        Deep-SHC
                                    label : Deep-SHC
                                 mgmt_uri : https://172.31.72.220:8089
                           mgmt_uri_alias : https://172.31.72.220:8089
                                   status : Up
```

Figure 18-9. Configuring search head clustering captain node

To check the status of a search head cluster member using a captain, refer to the following command:

```
sudo ./splunk show shcluster-status
```

This section implemented multisite search head clustering for site 1. In the next section, you use deployment instances to deploy configuration files in Splunk.

Deploying Configurations

In this section, you would deploy a configuration for the following task:

- Deploy indexes.conf and props.conf to all indexers using the cluster manager.

- Deploy shc app to the search head clustering using the deployer.

- Configure the cluster manager and the universal forwarder for indexer discovery.

- Deploy configuration on universalforwarder1 using the deployment server to monitor the Test.txt file.

- Configure the manager node and deployment server to send internal logs to the indexers.

Configuring a Cluster Manager

The manager node can distribute configuration to indexers. In the use case, the universal forwarder will forward Test.txt data to the index test.

1. To deploy a configuration using a manager node, go to `$SPLUNK_HOME/etc/manager-apps/_cluster/local`.

2. Create or edit the indexes.conf file. Refer to the following code section:

```
[test]
homePath=$SPLUNK_DB/test/db
coldPath=$SPLUNK_DB/test/colddb
thawedPath=$SPLUNK_DB/test/thaweddb
repFactor=auto
```

3. Create or edit the props.conf file for source type
    Test9. Refer to the following code section:

```
[Test9]
TIME_PREFIX=\d{1,3}\.\d{1,3}\.\d{1,3}\.\d{1,3}\s\-
\s\d{5}\s+
TIME_FORMAT = %m/%d/%Y %k:%M
MAX_TIMESTAMP_LOOKAHEAD = 15
LINE_BREAKER = ([\r\n]+)\d+\s+\"\$EIT\,
SHOULD_LINEMERGE = false
```

4. To deploy a configuration file to all indexers using a
    cluster manager, go to $SPLUNK_HOME/bin/.

```
$ sudo ./splunk validate cluster-bundle
$ sudo ./splunk apply cluster-bundle
```

5. To confirm the status of the bundle validation, refer
    to the following command:

```
$ sudo ./splunk show cluster-bundle-status
```

If your cluster bundle deploys successfully, you find the checksum of
your deployment. The command is similar to Figure 18-10.

```
[ec2-user@ip-172-31-75-109 bin]$ sudo ./splunk validate cluster-bundle  --check restart
Validating new bundle. Please run 'splunk show cluster-bundle-status' to check the status of the bundle validation.
[ec2-user@ip-172-31-75-109 bin]$ sudo ./splunk show cluster-bundle-status

master
        cluster_status=None
        active_bundle
                checksum=826AF3010CA7165419661216C27A2AAD
                timestamp=1592876799 (in localtime=Tue Jun 23 01:46:39 2020)
        latest_bundle
                checksum=826AF3010CA7165419661216C27A2AAD
                timestamp=1592876799 (in localtime=Tue Jun 23 01:46:39 2020)
        last_validated_bundle
                checksum=826AF3010CA7165419661216C27A2AAD
                last_validation_succeeded=1
                timestamp=1592918946 (in localtime=Tue Jun 23 13:29:06 2020)
        last_check_restart_bundle
                last_check_restart_result=restart not required
                checksum=
                timestamp=0 (in localtime=Thu Jan  1 00:00:00 1970)

  idx3    0700F4C5-9908-4036-AE0C-58A347989BAA     site2
          active_bundle=826AF3010CA7165419661216C27A2AAD
          latest_bundle=826AF3010CA7165419661216C27A2AAD
          last_validated_bundle=826AF3010CA7165419661216C27A2AAD
          last_bundle_validation_status=success
          restart_required_apply_bundle=0
          status=Up

  idx4    DC29F3C9-32C4-454C-8D53-8CDCAA618549     site2
          active_bundle=826AF3010CA7165419661216C27A2AAD
          latest_bundle=826AF3010CA7165419661216C27A2AAD
          last_validated_bundle=826AF3010CA7165419661216C27A2AAD
          last_bundle_validation_status=success
          restart_required_apply_bundle=0
          status=Up

  idx1    DCD95AC2-E49C-4CEB-A9D4-C54B00C2B832     site1
          active_bundle=826AF3010CA7165419661216C27A2AAD
          latest_bundle=826AF3010CA7165419661216C27A2AAD
          last_validated_bundle=826AF3010CA7165419661216C27A2AAD
          last_bundle_validation_status=success
          restart_required_apply_bundle=0
          status=Up

  idx2    FEB0AD2C-4FEF-43DC-83D1-3230BFCD80EE     site1
          active_bundle=826AF3010CA7165419661216C27A2AAD
          latest_bundle=826AF3010CA7165419661216C27A2AAD
          last_validated_bundle=826AF3010CA7165419661216C27A2AAD
          last_bundle_validation_status=success
          restart_required_apply_bundle=0
          status=Up
```

Figure 18-10. *Deploying cluster bundle on peer indexers*

Deploying an App to Search Head Cluster Using the Deployer

In this section, as provided in the use case to make a scalable platform, you do not need to push all configurations by going to the node, so you use the deployer to push configurations. The shc app sends internal logs to indexers.

To deploy the shc app using a deployer, go to $SPLUNK_HOME/bin/ in all the shs and run the following command:

```
$ sudo ./splunk edit shcluster-config -conf_deploy_fetch_url
https://172.31.77.218:8089
```

In the deployer, create an app named shc located in $SPLUNK_HOME /etc/shcluster/apps/.

1. Create or edit app.conf located in /local/app.conf in the shc app. Refer to the following code:

```
[ui]
is_visible = 0

[package]
id = shc_base
check_for_updates = 0
```

2. Create or edit outputs.conf located in /local/outputs.conf for the shc app. Refer to the following code:

```
[indexAndForward]
index = false

[tcpout]
defaultGroup = default-autolb-group
forwardedindex.filter.disable = true
indexAndForward = false

[tcpout:default-autolb-group]
server=172.31.75.49:9997,172.31.68.169:
9997,172.31.33.226:9997,172.31.42.237:9997
```

3. To push the configuration to the search head cluster nodes, go to $SPLUNK_HOME/bin/ in the deployer and run the following command:

```
$ sudo ./splunk apply shcluster-bundle \
-target https://<shc_member>:8089 \
-auth <user>:<password>
--answer-yes
```

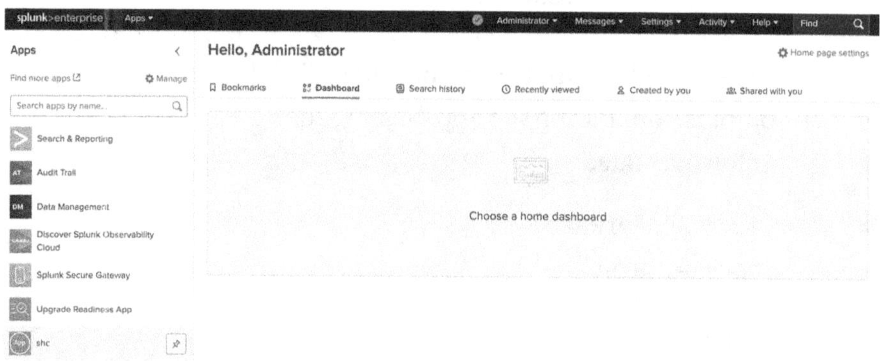

Figure 18-11. *Deploying cluster bundle on the search head using the search head cluster captain*

Figure 18-12. *SHC app deployed*

Configuring a Universal Forwarder for Indexer Deployment

In this section, you will deploy two apps using the deployment server:

1) An app to enable indexer discovery on universal forwarders.

2) The uf1 app that monitors the Test.txt file will be deployed on Universal Forwarder 1.

Configuring a Cluster Manager for Indexer Discovery

To configure indexer discovery in the cluster manager, you need to edit its server.conf.

1. Create or edit server.conf located in $SPLUNK_HOME/ etc/system/local/. Refer to the following code:

```
[indexer_discovery]
pass4SymmKey = my_secret
```

2. Restart the manager node by going to $SPLUNK_ HOME/bin/. Refer to the following command:

```
$ sudo ./splunk restart
```

Configuring a Forwarder for the Deployment Server

To configure forwarders for the deployment server, go to the EC2 instance of the forwarders, and migrate to $SPLUNK_HOME/bin/.

To configure the forwarders to report to the deployment server, refer to the following command:

```
sudo ./splunk set deploy-poll 172.31.77.218:8089
```

Configuring a Custom App for Indexer Discovery

To configure indexer discovery in a universal forwarder, you need to edit outputs.conf.

1. Create a new app in the deployment server, called
    fw_outputs, and add the outputs.conf in the
    $SPLUNK_HOME/etc/deployment-apps/fw_outputs/
    local. Refer to the following code:

    ```
    [indexer_discovery:manager]
    pass4SymmKey = my_secret
    manager_uri = https://172.31.75.109:8089

    [tcpout:group1]
    autoLBFrequency = 30
    autoLB = true
    indexerDiscovery = manager
    useACK=true
    ```

2. Create server.conf located in the $SPLUNK_HOME/
    etc/deployment-apps/fw_outputs/local. Refer to
    the following code:

    ```
    [general]
    site = site1

    [clustering]
    multisite = true
    ```

Configuring a Custom App for Monitoring

Configure the deployment server to serve the uf1 app following this process:

1. Create an app named uf1 in $SPLUNK_HOME/etc/deployment-apps/.

2. Go to uf1/local, and create inputs.conf to monitor the Test.txt file. Refer to the following stanza:

```
[monitor:///opt/Test.txt]
disabled=false
index=test
sourcetype=Test9
```

Deploying the UF1 App

Use forwarder management to deploy the uf1 app on a universal forwarder.

1. Go to the deployment server on Splunk Web.

2. Go to Settings in Splunk Web and then to Agent Management.

Two universal forwarders are reporting to the agent management. Refer to Figure 18-13.

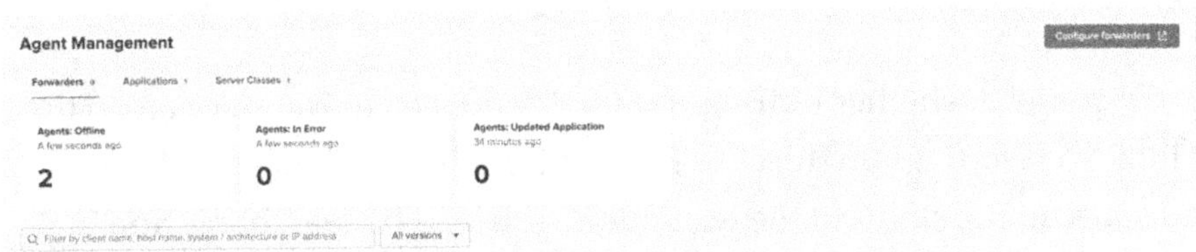

Figure 18-13. *Universal forwarders in Agent Management*

3. Once universal forwarder 1 reaches out to the deployment server, you can create two new server classes:

    - **uf_inputs**: Add the uf1 app and the universal forwarder 1 as an agent.

- **uf_outputs**: Add the fw_outputs and both universal forwarders as agents.

Once the uf1 and fw_outputs apps are deployed in the universal forwarder 1, it will start sending its logs to the indexers. When you log in to a sh, and when you type index ="test" in the search bar, you can see the data is parsed from host= "DEEP-FORWARDER1". Refer to Figure 18-14.

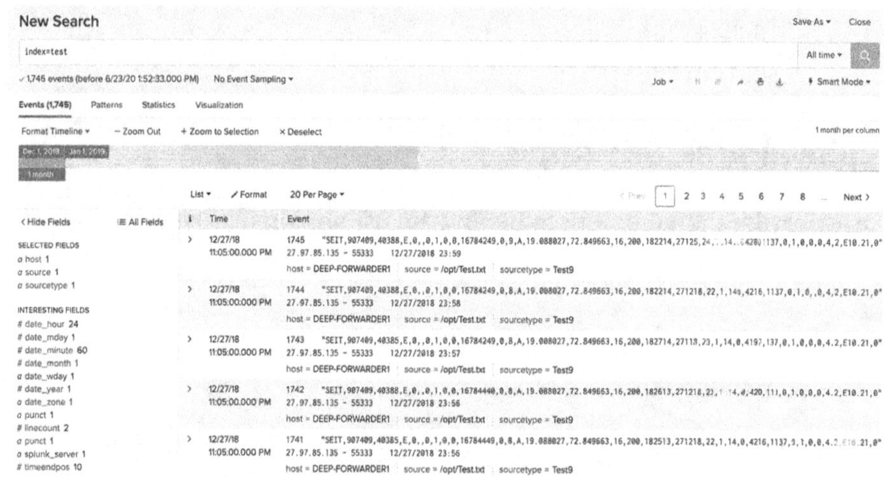

Figure 18-14. *Events in the index test for DEEP-FORWARDER1*

To confirm whether repFactor=auto for the test index, navigate to the Splunk Web settings on the cluster manager and go to Indexer Clustering. You can find the status of the test index. It should be searchable and should have replicated data copies. Refer to Figure 18-15.

⊘ All Data is Searchable		⊘ Search Factor is Met		⊘ Replication Factor is Met

4 searchheads 0 not searchable	4 searchheads 0 not searchable
Peers	Indexes

Peers (4)   Indexes (4)   Search Heads (1)

Bucket Status ▾   Find...   🔍   10 per page ▾

Index Name ⬍	Fully Searchable ⬍	Searchable Data Copies ⬍	Replicated Data Copies ⬍	Buckets ⬍	Cumulative Raw Data Size ⬍
_audit	✓ Yes	2	2	4	< 0.01 GB
_internal	✓ Yes	2	2	4	< 0.01 GB
_telemetry	✓ Yes	2	2	4	< 0.01 GB
test	✓ Yes	2	2	2	< 0.01 GB

Figure 18-15. Indexer Clustering

Configuring the Manager Node, Deployment Server, and Search Head 4 to Send Internal Logs

In this section, you send internal logs from the deployment server, manager node, and search head 4 to the indexers.

1. Go to the deployment server, cluster manager, and search head 4 instances.

2. Go to $SPLUNK_HOME/etc/system/local/outputs. conf. Refer to the following code:

```
[indexer_discovery:manager]
pass4SymmKey = my_secret
manager_uri  = https://172.31.75.109:8089

[tcpout]
defaultGroup = group1
autoLB = true
forwardedindex.filter.disable = true
indexAndForward = false
```

```
[tcpout:group1]
indexerDiscovery = manager
autoLB = true
autoLBFrequency = 30
useACK = true
```

3. Go to $SPLUNK_HOME/bin/, and type the following command:

```
$ sudo ./splunk restart
```

Now let's look at monitoring distributed environments.

Monitoring Distributed Environments

In this section, you use a Monitoring Console to monitor distributed environments. The Monitoring Console monitors distributed environments. It troubleshoots when a Splunk instance fails or when any other issue occurs in your Splunk environment.

Adding a Search Peer to Monitor

In this section, you monitor your Splunk environment using the Monitoring Console in Amazon Web Services EC2 instances. To monitor the Splunk environment, you need to add the deployment's instances as search peers in the Monitoring Console. To add a search peer, refer to the following instructions:

1. Using Splunk Web, go to Monitoring Console Instance, Settings, Distributed search, and Search peers.

2. Click New Search Peer, and enter the peer URI, followed by the remote username and remote password. Confirm the password.

3. Click Save.

Once you have added all the search peers to your instance, you see a screen similar to Figure 18-16.

Search peers

Distributed search > Search peers

Showing 1-9 of 9 items

Peer URI ⍗	Splunk instance name ⍗	State ⍗	Replication status ⍗	Cluster label ⍗	Health status	Health check failures	Status ⍗	Actions
172.31.33.226:8089	idx3	Up	Initial	5212C962-77E5-416B-8BE3-CE030286C451	✓ Healthy	None	Enabled I Disable	Quarantine I Delete
172.31.42.237:8089	idx4	Up	Initial	5212C962-77E5-416B-8BE3-CE030286C451	✓ Healthy	None	Enabled I Disable	Quarantine I Delete
172.31.65.106:8089	sh2	Up	Initial	5212C962-77E5-416B-8BE3-CE030286C451	✓ Healthy	None	Enabled I Disable	Quarantine I Delete
172.31.68.169:8089	idx2	Up	Initial	5212C962-77E5-416B-8BE3-CE030286C451	✓ Healthy	None	Enabled I Disable	Quarantine I Delete
172.31.70.85:8089	sh4	Up	⊘ Successful	5212C962-77E5-416B-8BE3-CE030286C451	✓ Healthy	None	Enabled I Disable	Quarantine I Delete
172.31.72.220:8089	Deep-SHC	Up	⊘ Successful	5212C962-77E5-416B-8BE3-CE030286C451	✓ Healthy	None	Enabled I Disable	Quarantine I Delete
172.31.74.232:8089	sh1	Up	⊘ Successful	5212C962-77E5-416B-8BE3-CE030286C451	✓ Healthy	None	Enabled I Disable	Quarantine I Delete
172.31.75.49:8089	idx1	Up	Initial	5212C962-77E5-416B-8BE3-CE030286C451	✓ Healthy	None	Enabled I Disable	Quarantine I Delete
172.31.75.109:8089	cmaster	Up	⊘ Successful	5212C962-77E5-416B-8BE3-CE030286C451	✓ Healthy	None	Enabled I Disable	Quarantine I Delete

Figure 18-16. Search peers in the monitoring console

General Setup for Distributed Environments

To set up a general distributed environment in Splunk, go to the Monitoring Console instance in Splunk Web and refer to the following instructions:

1. Go to the Monitoring Console and then to Settings in Splunk Web.

2. Go to Monitoring Console.

3. Click Settings and then go to General Setup.

4. Select Distributed under the Mode option. Click Continue. You should see the instances.

5.   Examine the auto-selected roles. Edit the roles that are not configured according to their use case.

6.   After editing all roles according to their instance, click Apply Changes.

7.   Save all configured changes, and go to the Overview page.

On the Overview page, you find the deployment diagram shown in Figure 18-17.

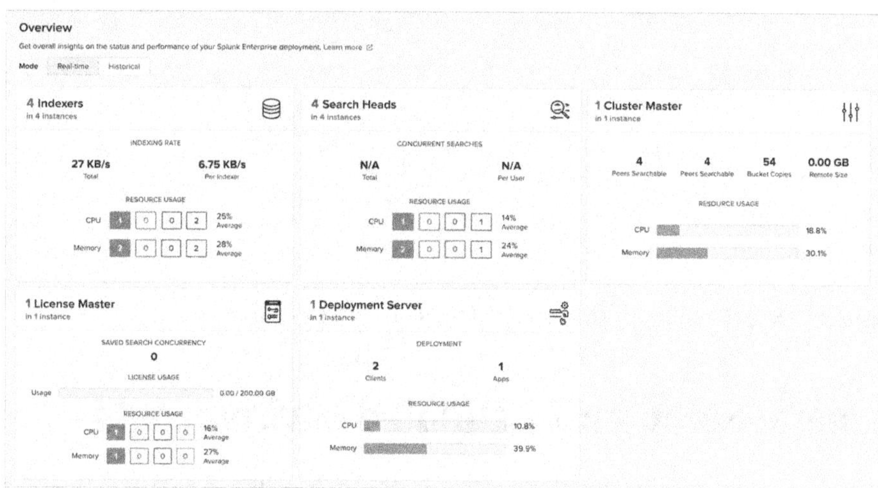

Figure 18-17. *Monitor console overview*

Conclusion

In this chapter, you learned how to deploy a Splunk environment on the AWS platform and how the core building blocks fit together in a real cloud setup.

That brings us to the end of this book. Our goal was to give you a clear, practical path through the topics that matter most for the Splunk Administrator certification, from inputs and parsing to indexing, clustering, and deployment at scale. Of course, Splunk is a big ecosystem, and the learning doesn't stop here. When you want to go deeper, the official Splunk Documentation is the best source of truth, and it's where you'll find the latest details, examples, and edge-case behaviors. When you hit something weird in the real world (because you will), the Splunk Community is one of the strongest parts of the platform since thousands of admins, engineers, and architects share solutions, ideas, and hard-earned lessons.

Most importantly, don't treat this as a finish line; treat it as your launch pad. Keep building labs, keep breaking things on purpose, keep reading logs like a detective, and keep asking "why" until the system makes sense in your (search) head. You've got the foundation now, go use it, trust your process, and enjoy the work. Happy Splunking, and good luck on your exam and beyond!

References

- https://help.splunk.com/en/splunk-enterprise/
  administer/manage-indexers-and-indexer-
  clusters/10.0/deploy-and-configure-a-multisite-
  indexer-cluster/configure-multisite-indexer-
  clusters-with-the-cli

- https://help.splunk.com/en/splunk-enterprise/
  administer/distributed-search/10.0/configure-
  search-head-clustering/configure-the-search-
  head-cluster

- https://help.splunk.com/en/splunk-enterprise/
  administer/manage-indexers-and-indexer-
  clusters/10.0/manage-indexes/create-
  custom-indexes

- https://help.splunk.com/en/splunk-enterprise/
  administer/manage-indexers-and-indexer-
  clusters/10.0/get-data-into-the-indexer-
  cluster/use-indexer-discovery-to-connect-
  forwarders-to-peer-nodes

© Carlos Moreno Buitrago, Deep Mehta 2026

C. M. Buitrago and D. Mehta, *The Splunk Core User Study Companion*, Certification Study Companion Series, https://doi.org/10.1007/979-8-8688-2501-9

B

C

D

W, X, Y, Z

GPSR Compliance
The European Union's (EU) General Product Safety Regulation (GPSR) is a set
of rules that requires consumer products to be safe and our obligations to
ensure this.

If you have any concerns about our products, you can contact us on

ProductSafety@springernature.com

In case Publisher is established outside the EU, the EU authorized
representative is:

Springer Nature Customer Service Center GmbH
Europaplatz 3
69115 Heidelberg, Germany